AUTODESK® INVENTOR® 2014
AND INVENTOR LT™ 2014

ESSENTIALS

AUTODESK® INVENTOR® 2014
AND **INVENTOR LT™ 2014**

ESSENTIALS

Thom Tremblay

AUTODESK®
Official Press

SYBEX®
A Wiley Brand

Senior Acquisitions Editor: Willem Knibbe
Development Editor: Alexa Murphy
Technical Editor: Dan Hunsucker
Production Editor: Eric Charbonneau
Copy Editor: Liz Welch
Editorial Manager: Pete Gaughan
Production Manager: Tim Tate
Vice President and Executive Group Publisher: Richard Swadley
Vice President and Publisher: Neil Edde
Book Designer: Happenstance Type O Rama
Proofreader: Nancy Bell
Indexer: Ted Laux
Project Coordinator, Cover: Katherine Crocker
Cover Designer: Ryan Sneed
Cover Image: © Thom Tremblay

ISBN: 978-1-118-57520-8
ISBN: 978-1-118-75759-8 (ebk.)
ISBN: 978-1-118-75757-4 (ebk.)

For general information on our other products and services or to obtain technical support, please contact our Customer Care Department within the U.S. at (877) 762-2974, outside the U.S. at (317) 572-3993 or fax (317) 572-4002.

Wiley publishes in a variety of print and electronic formats and by print-on-demand. Some material included with standard print versions of this book may not be included in e-books or in print-on-demand. If this book refers to media such as a CD or DVD that is not included in the version you purchased, you may download this material at http://booksupport.wiley.com. For more information about Wiley products, visit www.wiley.com.

Library of Congress Control Number: 2013936845

Dear Reader,

Thank you for choosing *Autodesk Inventor 2014 and Inventor LT 2014 Essentials*. This book is part of a family of premium-quality Sybex books, all of which are written by outstanding authors who combine practical experience with a gift for teaching.

Sybex was founded in 1976. More than 30 years later, we're still committed to producing consistently exceptional books. With each of our titles, we're working hard to set a new standard for the industry. From the paper we print on to the authors we work with, our goal is to bring you the best books available.

I hope you see all that reflected in these pages. I'd be very interested to hear your comments and get your feedback on how we're doing. Feel free to let me know what you think about this or any other Sybex book by sending me an email at nedde@wiley.com. If you think you've found a technical error in this book, please visit http://sybex.custhelp.com. Customer feedback is critical to our efforts at Sybex.

Best regards,

NEIL EDDE
Vice President and Publisher
Sybex, an Imprint of Wiley

To the code writers, product developers, product managers, and program managers who created Inventor and made a difference: thank you.

Acknowledgments

I want to thank the tremendous team at Sybex for their patience and professionalism; specifically, Willem Knibbe, Pete Gaughan, Alexa Murphy, Eric Charbonneau, Liz Welch, Jenni Housh, Rebekah Worthman, Connor O'Brien, Rayna Erlick, and everyone else who worked hard behind the scenes with whom I didn't get a chance to communicate directly. Special thanks once again to Dan Hunsucker for being the technical editor. If you're in the Kansas City area and want to learn to use the Autodesk® Inventor® software from a real expert, you'll be in great hands with Dan. Thanks to Tom Joseph, Don Carlson, and Debra Pothier of Autodesk for their support. Most of all, thank you to my wife Nancy for tolerating another "book season."

ABOUT THE AUTHOR

Thom Tremblay is a Strategy Manager for Autodesk Education and has worked with hundreds of schools and companies to help them understand how Inventor can help them with their designs. He is an Autodesk Inventor Certified Professional, an Autodesk Certified Instructor, and an Autodesk Certification Evaluator, and he has been working with Inventor for more than 14 years and with other Autodesk products for more than 27 years. He has used Autodesk software to design everything from cabinets and castings to ships and video monitors. He has close ties to the Inventor community; is a frequent speaker at colleges, universities, and training centers; and presents at Autodesk University annually.

CONTENTS AT A GLANCE

Contents

CHAPTER 11　Working with Sheet Metal Parts　291

CHAPTER 12　Building with the Frame Generator　321

INTRODUCTION

This book is designed to be a direct, hands-on guide to learning Inventor by *using Inventor*. The book includes lessons for absolute beginners, but experienced users can also find exercises to show them how tools they're not familiar with work.

Nearly all of the over 200 exercises can be started from an existing file, so you need to do only those exercises that will help you most.

Who Should Read This Book

Autodesk Inventor 2014 and Inventor LT 2014 Essentials is designed to meet the needs of the following groups of users:

- ▶ Professionals who use 2D or 3D design systems and want to learn Inventor at their own pace

- ▶ Professionals attending instructor-led Inventor training at an Autodesk Authorized Training Center

- ▶ Engineering and design students who need to learn Inventor to support their education and career

What You Will Learn

Autodesk Inventor 2014 and Inventor LT 2014 Essentials covers the most common uses of the tools in Autodesk Inventor® and Inventor LT®. Not every option is covered, but you will easily understand the options once you learn to use the primary tool.

The first nine chapters cover the core of Inventor in a way that steps the reader through creating drawings, parts, and assemblies in phases so there is a better opportunity to absorb the concepts.

The second portion of the book consists of four chapters that focus on tools and workflows specific to types of design and visualization. I recommend readers complete the work in these chapters to learn alternative workflows that may not be an obvious fit for their design needs but may help them nonetheless.

What You Need

To perform the exercises in this book, you must have Autodesk Inventor 2014 or Inventor LT 2014.

FREE AUTODESK SOFTWARE FOR STUDENTS AND EDUCATORS

The Autodesk Education Community is an online resource with more than 5 million members that enables educators and students to download — for free (see the website for terms and conditions) — the same software used by professionals worldwide. You can also access additional tools and materials to help you design, visualize, and simulate ideas. Connect with other learners to stay current with the latest industry trends and get the most out of your designs. Get started today at www.autodesk.com/joinedu.

While offering a somewhat limited feature set, Inventor LT is a flexible and powerful tool focused on the core user needs of part modeling and 2D drawings. Inventor LT does not use a project file.

Autodesk Inventor LT is best aligned with Chapters 1–4 and 6–8 but is capable of doing some of the functions covered in Chapters 10 and 11. In cases where an assembly is used for an exercise on visualization you can use a part file instead.

To make sure that your computer is compatible with Autodesk Inventor 2014, check the latest hardware requirements at www.autodesk.com/inventor.

What Is Covered in This Book

Autodesk Inventor 2014 and Inventor LT 2014 Essentials is organized to provide you with the knowledge needed to master the basics of Inventor.

Chapter 1: Connecting to the Interface   This chapter presents the interface, the basics of working with Inventor, and how to become productive with Inventor.

Chapter 2: Introducing Parametric Sketching   The fundamental element of a 3D part is a 2D sketch using parametric dimensions to control its size. This chapter shows how to create these sketches.

Chapter 3: Introducing Part Modeling   Building parametric solid models is essential to the effective use of Inventor. This chapter introduces the tools you need to build basic parts in Inventor.

Chapter 4: Creating 2D Drawings from 3D Data   Creating 2D documentation of your designs is critical. This chapter presents the basic tools for placing views and dimensions in your drawings.

Chapter 5: Introducing Assembly Modeling   Most products are made of many parts. Assembly tools help you control the position of the components relative to one another.

Chapter 6: Exploring Part Modeling   In this chapter you dive deeply into the structure of 3D part models and learn techniques for refining your control over features.

Chapter 7: Advanced Part Modeling   Advanced geometry requires more advanced modeling tools. Learn how to use advanced fillets, lofts, and other tools that create the complex shapes you need.

Chapter 8: Creating Advanced Drawings and Annotations   This chapter focuses on creating and editing more complex drawing views and adding finishing touches to your drawings.

Chapter 9: Advanced Assembly and Engineering Tools   An assembly is more than a group of parts. Inventor features many engineering-based tools that work in the assembly. This chapter also describes tools to help you control complex assemblies.

Chapter 10: Creating Sculpted and Multibody Parts   Plastics have a number of common features that make them easier to assemble. These features are developed using specialized tools in Inventor.

Chapter 11: Working with Sheet Metal Parts   The process of manufacturing sheet metal parts heavily influences how they are designed in Inventor. Creating material styles makes it easy for you to change the components by changing the style.

Chapter 12: Building with the Frame Generator   Using traditional solid modeling tools to build metal frames is arduous and time-consuming. The Frame Generator tools shortcut the process and make even complex frames easy to design.

Chapter 13: Working in a Weldment Environment   A weldment is a combination of an assembly and a part model. Inventor puts the needs of manufacturing first when defining a weldment, saving you time.

Appendix: Autodesk Inventor Certification   Show the world that you know Inventor by becoming an Autodesk Certified User, Associate, or Expert. This appendix will help you find the resources in the book to get certified.

Visit the book's website at www.sybex.com/go/inventor2014essentials for this additional content:

Bonus Chapter 1: Customizing Styles and Templates   Using standards in manufacturing improves quality and efficiency. The same is true for Inventor. This chapter helps you understand the options that are available for building your own design standard.

Bonus Chapter 2: Working with Non-Inventor Data   Inventor has the ability to import and export data to and from nearly any other design system. This chapter helps you understand what options you have in working with that data.

Bonus Chapter 3: Automating the Design Process and Table-Driven Design   If you have repeatable design processes and products that share a lot of common features using tables can greatly improve productivity. Using rules to create new components that comply with production capabilities gives you even more flexibility.

Bonus Chapter 4: Creating Images and Animation from Your Design Data   Sharing images and animations made from your designs can help others understand how your designs are created and understand their value. This chapter guides you through tools for sharing your work with others.

Exercise Data   To complete the exercises in *Autodesk Inventor 2014 and Inventor LT 2014 Essentials*, you must download the data files from www.sybex.com/go/inventor2014essentials.

Please also check the book's website for any updates to this book should the need arise. You can also contact the author directly by email at inventor.essentials@outlook.com.

The Essentials Series

The Essentials series from Sybex provides outstanding instruction for readers who are just beginning to develop their professional skills. Every Essentials book includes these features:

- ► Skill-based instruction with chapters organized around projects rather than abstract concepts or subjects.

- ► Suggestions for additional exercises at the end of each chapter, where you can practice and extend your skills.

- ► Digital files (via download) so you can work through the project tutorials yourself. Please check the book's web page at www.sybex.com/go/inventor2014essentials for these companion downloads.

CHAPTER 1

Connecting to the Interface

To access the power of Autodesk® Inventor® 2014, you have to start with the interface. To some extent, Inventor is an interface between your ideas and your computer.

The ability to navigate and leverage the nuances of a program interface can mean the difference between struggling and excelling with the application. In this chapter, you will explore the components of dialog boxes, Ribbons, tabs, and viewing tools that will help you create your designs. You will also learn how to modify the interface to increase your comfort with Inventor.

▶ **Exploring the graphical user interface**

▶ **Setting application options**

▶ **Using visualization tools**

▶ **Working with project files**

Exploring the Graphical User Interface

Autodesk Inventor 2014 opens with a Welcome screen (Figure 1.1), which presents the new user with the tools needed to create a new file, open an existing one, and access the many Essential Skills Videos and tutorials built into the product. You can even change the units of the default templates to spare you from selecting Metric or English templates all the time.

FIGURE 1.1   The Autodesk Inventor Welcome screen

When you first see the interface, you will probably think it is rather bare. With no file open, you just have the absolute basics there. Even when a file is loaded, your design remains the focus of the interface. In Figure 1.2, you can see the primary elements of the interface that we will refer to in this chapter.

Users of other current Autodesk or Microsoft applications will recognize the Ribbon-style interface and the Application icon in the upper left. The adoption of the Ribbon interface in Inventor goes beyond most other applications by actively offering you tools when they're most needed. But let's not get ahead of ourselves; let's start by getting more details on these features.

Across the top of the Inventor window is the title bar. It lets you know you're using Autodesk Inventor, or it displays the name of the active file when you're editing one.

In the upper-left corner is an icon with a large *I* on it. Clicking it opens the Application menu (Figure 1.3), which displays tools for creating and manipulating files (on the left) and a list of recently opened files (on the right). If you want to be able to return to a file frequently, you can select the pushpin icon to the right of the filename and keep it on the list of recently accessed files.

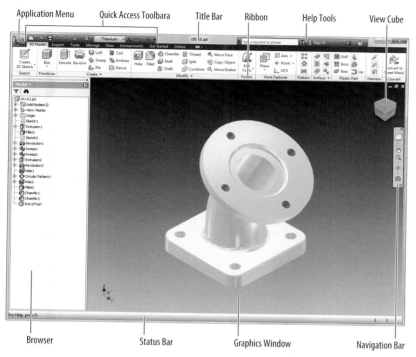

FIGURE 1.2 Elements of the Inventor user interface

You can also toggle the list between recent documents and the documents that are currently open and change the list from filenames to icons showing the files.

At the top (on the right side) is a *Search Commands* window for finding tools that are available in the file you are editing. As you begin typing a tool name or more conceptual term, a list of options based on the entry will be generated. You can then start the tool from that list. The location of the tool in the Ribbon is also displayed for future reference.

Just below that you will find icons that sort how previously opened files are listed and a drop-down menu for the size of the icon displayed.

At the bottom of the menu are buttons to exit Inventor and to access the application options, which you will explore later in this chapter.

The Quick Access toolbar is embedded in the title bar next to the Application icon and contains common tools for accessing new file templates, undoing and redoing edits, and printing. The toolbar is dynamic, which means different tools will appear depending on the active file. For example, one of those part-time tools is a drop-down menu that allows you to change the color of the active part.

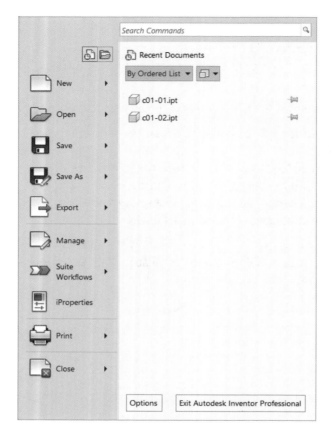

FIGURE 1.3 The expanded Application menu and
Quick Access toolbar

You can customize this toolbar by adding commonly used tools to the toolbar.
To do so, select the desired tool from the Ribbon, right-click, and select Add To
Quick Access Toolbar from the context menu.

IMPORTANT!

At this time, it is critical that you go to www.sybex.com/go/inventor2014
essentials, download the Inventor 2014 Essentials Data.zip
file, and extract it to the root of your C drive. This will create a new folder,
C:\Inventor 2014 Essentials. The files needed to complete the
exercises for this book are in this folder.

Opening a File

Knowing this much of the interface will allow you to access the Open dialog box and see how the rest of the interface works. In this exercise, you will open a file in Inventor:

1. Start Autodesk Inventor if it is not already up and running.

2. Expand the Application menu, and select Open from the options on the left.

   When the Open dialog box appears, notice the tools at the top. These tools allow you to navigate to other folders as you would in Windows Explorer, change the way files are displayed (including the option of thumbnail images), and add new folders.

3. In the Open dialog box, use the Up One Level button to navigate to the C:\Inventor 2014 Essentials\Parts\Chapter 01, folder, as shown in Figure 1.4.

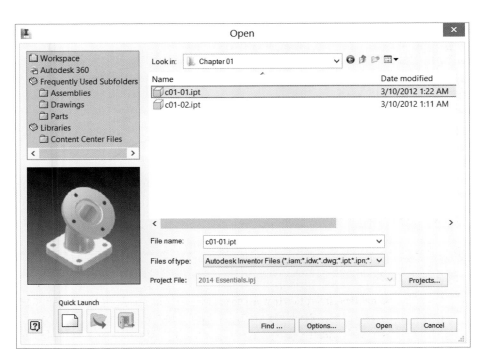

FIGURE 1.4 The Open dialog box includes tools for finding and browsing for files.

You can also access the Open tool via the Quick Access toolbar.

Selecting an Inventor file from the file list generates an image in the Preview pane on the left side of the dialog box.

Certification Objective

4. Double-click the c01-01.ipt file listed, or click it once and then click Open.

5. To see the complete model, move your cursor near the ViewCube® in the upper right of the Graphics window. When the icon that looks like a house appears, click it.

Clicking the Home view icon restores a view position that was saved with the model. Your screen should now resemble Figure 1.2.

KNOW YOUR ICONS!

These are the icons for the four primary Inventor file types:

.idw for 2D drawings

.ipt for 3D parts

.iam for 3D assemblies

.ipn for 3D presentation files

The Open dialog box, templates, and even Windows Explorer use these icons as a visual reference for the type of information they contain.

Exploring the Ribbon

You will now see that the Ribbon has more options available. These options take the form of tools that are grouped into panels on tabs. Let's look at these in order from the broadest to the most specific:

Tabs   The options shown as 3D Model, Inspect, Tools, and so on are referred to as *tabs*. The active tab always contrasts with the others that just appear as names on the background of the Ribbon. The active tabs can change automatically as you transition from one working environment to another.

There are tabs that appear temporarily when you use specialized tools or enter a specialized environment, such as sketching (Figure 1.5) or rendering a 3D model. These contextual tabs activate automatically and have a special green highlight to help them stand out and remind you that the tools are available.

FIGURE 1.5   The Sketch tab is not normally displayed but gets special highlighting when it is.

Panels   On each tab are special collections of tools that are sorted into *panels*. For example, in Figure 1.5, you can see Draw, Constrain, Pattern, and other panels on the Sketch tab. Notice that the most commonly used tools appear with larger icons so they're easier to locate. At times, not all of the tools in a panel can fit, so some of the panels (Draw and Constrain) have down arrows next to their names. This indicates that you can expand the panel to see more tools. It is also possible to select tools that you don't often use and place them permanently in this expandable portion of the panel. You can even rearrange the order of the panels in a tab.

Tools   The last element of the Ribbon is the icon or tool. Many of the tool names display near the icon. As you become more comfortable, you can save space on your screen by turning off the display of the names. You can also hide or partially hide the Ribbon if you prefer having a larger Graphics window.

Any time you see a down arrow next to a word or icon in Inventor, it means there are additional options.

   Let's try a few of these options so that you can make yourself more comfortable once you start using Inventor for your regular work:

    1. Keep using or reopen the c01-01.ipt file used in the previous exercise.

To the right of the tab headers is a gray icon with an up arrow. Click this arrow to cycle the various ways of displaying the Ribbon.

2. Click the icon once to change the Ribbon from displaying all the panels within a tab to showing only the first icon with the title of the panel below it, as in Figure 1.6.

FIGURE 1.6 The 3D Model tab showing Panel Buttons mode

When you hover over or click the icons that represent a panel (or the panel or tab titles of the next steps), a full view of the panel appears under the Ribbon.

3. Click the same icon a second time, and you will reduce the tab to showing only the title of the panels.

4. Now, click the icon again to reduce the Ribbon even further to only display the names of the tabs.

5. Click the icon one last time to restore the Ribbon to its original size with the default panels displayed.

Depending on your needs and personal preferences, you can use these various modes full-time or just temporarily. Clicking the down arrow to the right of the icon displays a menu that allows you to jump directly to the mode you want.

Rearranging the Panels

Let's do one more exercise displaying the Ribbon tools:

1. Keep using or reopen the c01-01.ipt file used in the previous exercise.

2. Click the 3D Model tab.
   The title bar of the panel can be used as a grip.

3. Click and drag the Pattern panel (Figure 1.7) so that it appears between the Create and Modify panels.

FIGURE 1.7 Any panel can be relocated to suit your tastes.

A panel can also be left floating in the interface to act as a traditional toolbar.

4. Right-click one of the panels, and expand the Ribbon Appearance menu by hovering over the words in the context menu.

5. Select Text Off from the menu.
   Doing so will remove the names of the tools in the panel. This cuts down on the screen area covered but keeps the tools readily available.

6. Right-click again in a panel to open a menu with Ribbon appearance options.

7. Select Reset Ribbon.

8. Click Yes when the warning dialog box appears.

After a few moments, the icon names reappear, and the panels revert to their original positions. This is a great safety net for experimenting with interface personalization.

Finally, note that hovering your cursor over a tool in the Ribbon will open a *tooltip* that will give you basic information, the correct name of the tool, and, when applicable, the keyboard shortcut for the tool. If you leave the cursor over that tool a little longer, the tooltip will expand to offer visual information about the use of the tool (Figure 1.8). Some tooltips even offer animated guidance on the tool.

The Help system in Inventor uses multiple resources to locate information for you. The Help tools are on the right end of the title bar, and they include a search window that will locate information from multiple resources, including the wiki for Inventor. You can even share your own tips on the wiki.

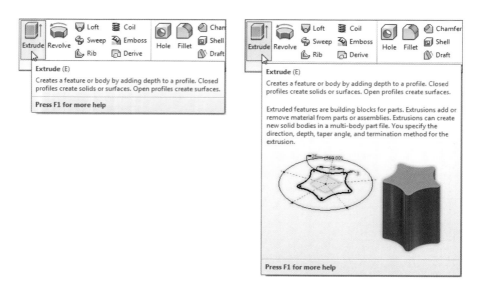

FIGURE 1.8   Continuing to hover expands the tooltip.

Using the Browser

The process of design in Inventor is often referred to as *parametric solid modeling.* Although parameters play an important role in using Inventor, what makes its design process powerful and flexible is its ability to maintain a history of how the parts were constructed. This history also shows how they are related to one another in an assembly. The interface to see these actions is the Browser.

On the left side of the screen is a column that contains a hierarchical tree showing the name of the active file at the top, the Representations and Origin folders, and then the components or pattern of components that make up your assembly. Understanding how to read the Browser and how to control it is an important step in understanding how to control and edit your designs.

Let's do some exploration of the Browser to see what it can tell you:

1. Keep using or reopen the c01-01.ipt file used in the previous exercise.

2. In the Browser, find the Origin folder, and click the plus sign to the left of the folder to expand the folder's contents.

3. Pass your mouse over the planes and axes displayed in the Browser, and see how those entities preview in the Graphics window to the right.

4. Right-click the XY Plane icon, and select Visibility from the context menu that appears.

The Browser displays the features that make up the active part (see Figure 1.9). Since it is a part file, the Ribbon has also changed, making the Model tab active for adding features to the part.

A filter tool at the top of the Browser offers ways to limit what information is shown in the Browser. The Show Extended Names option will include additional information on model features such as fillet radius or hole size.

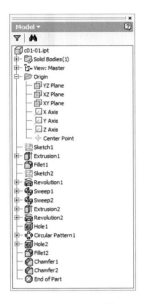

FIGURE 1.9 The Browser reflects the status of the part or assembly and its history.

If you activate a part in an assembly, it's also important to know how to return to the assembly view. On the right end of the Model tab, you will see the Return panel. Click the icon for the Return tool. This tool will appear at the end of a number of Ribbons when editing parts and assemblies.

In an assembly, there are additional options in the Return tool that you will use later in the book. Return's primary function is to take you up a step in the hierarchy of the model you're in.

Not only is Inventor capable of creating and editing parts while in the assembly, but it prefers to work that way. There is no need to open a part in a separate window to edit it, though at times you might find it easier to navigate. Once a part is activated, the inactive components in the assembly are grayed out in the Browser and fade in the Graphics window to make it easier to focus on the part you're editing.

You can also activate components in an assembly by double-clicking the geometry in the Browser or the Graphics window. Selecting a component and then right-clicking will allow you to open the part in its own window. This feature is not available in Autodesk® Inventor LT™.

Exploring the File Tabs

Once you've opened at least two files in Inventor, the file tabs appear. These tabs make it easier to navigate between the files and help you find the file that you want to switch to.

Let's open another part to see what other changes occur when you open a separate window to use for editing. Open the C:\Inventor 2014 Essentials\ Parts\Chapter 01\c01-02.ipt file in Inventor. In Figure 1.10, you can see the file tabs at the bottom of the Graphics window. These tabs display the names of the files.

FIGURE 1.10 File tabs let you quickly switch between open files and even preview the file before you select it.

The first two buttons on the left arrange your open files by cascading them or by displaying them as tiles side by side. Let's try using the tools to see what happens:

1. Click the Arrange button on the file tabs bar.
    The files will stack on top of each other, as shown in Figure 1.11.

2. Look at the Browser, and then click back and forth between the tabs for the c01-01.ipt and c01-02.ipt files to see the effect on the Browser.

3. Select the c01-02.ipt tab, and click the Close button (X) in the upper right of the tab.
    This closes the c01-02 part and the file tabs but leaves the other part open in a floating window.

4. Double-click the top of the floating window frame, or click the maximize icon near the upper-right corner of the window.
    This will restore the c01-01 part to filling the Graphics window.

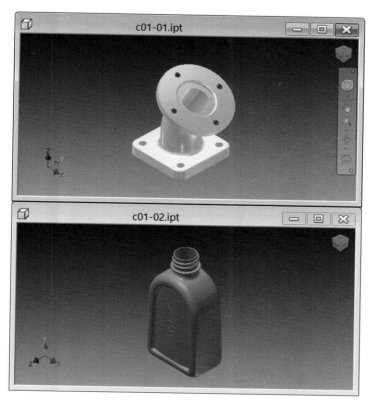

FIGURE 1.11 The files arranged as floating windows

The Graphics window has many different tools and enables many different options. I'll cover some of them in future chapters, but let's look at some of the most important ones now.

Highlighting and Enabled Components

In a part file, moving your cursor over faces and edges will highlight them for selection. The same is true for the views of a drawing. As you move your cursor over the components of an assembly in the Graphics window, you will notice that the parts highlight. Highlighting allows you to see when components stand alone or, as in the case of bolts in an assembly, as a group. When a component is highlighted in the Graphics window, it will also highlight in the Browser, and vice versa.

Sometimes, you want to visually reference an assembly component but do not want to be able to select that component or activate it. To do so, you can select the part in the Browser or the Graphics window and deselect Enabled from the context menu. This will fade the part, make it unselectable, and disable highlighting of the disabled part.

Working in the Graphics Window

Certification Objective

The part of the interface where most of your attention will be focused is the Graphics window. Not only is your design and its documentation shown there, but with each release of Inventor, more tools can be started from the Graphics window. Viewing and visualization tools can affect this area and allow you to view and share your designs in exciting ways.

Let's go through some of the tools that exist in and for the Graphics window and begin to see just how important it is:

Origin 3D Indicator   In the lower-left corner of the Graphics window, you see a 3D direction indicator. This is a reference for the orientation of the model in relationship to the standard axes of the file.

Pan   To slide the model or drawing around the Graphics window, use the Pan tool. You can access Pan in several ways. You can find the Pan tool by going to the Navigate panel on the View tab, by going to the Navigation bar in the Graphics window, by pressing the F2 key, or by clicking and holding the middle mouse button.

Zoom   The Zoom tool has several optional modes. The basic Zoom tool smoothly makes the model appear larger or smaller depending on the input you give it. You can find the Zoom tool by going to the drop-down under the Navigate panel on the View tab, by locating the Zoom icon on the Navigation bar, by pressing the F3 key and clicking and dragging your mouse button, or by rotating a mouse wheel.

Free Orbit   When you start Free Orbit by going to the Navigate panel or Navigation bar, or by pressing F4, a circle with lines at the quadrants sometimes referred to as a reticle appears on the screen. Moving your cursor around the reticle displays different tools. Figure 1.12 shows icons inside, outside, and near the x- and y-axes of the reticle. Clicking inside the reticle orbits the model in a tumbling manner. Clicking and dragging near the x- or y-axis or just outside of the reticle isolates the orbit around an axis.

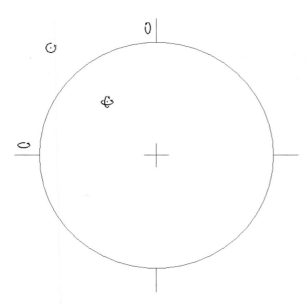

FIGURE 1.12 It is important to be able to easily reposition your view of the model for editing.

View History   You can return to a previous view or views by pressing and holding the F5 key until a view that you had before is restored on the screen.

ViewCube   It's very likely the current view of the assembly isn't the only useful view that you might need to develop the design. The ViewCube allows you to rotate your design and position it based on standard views aligned with the faces, edges, and corners of the cube. Rotating the model using the ViewCube works in a way similar to the Free Orbit tool.

The ViewCube is probably the most important tool in the Graphics window for keeping track of your design. In the next exercise, you will get a chance to begin using it.

Certification
Objective

1. Keep using or reopen the c01-01.ipt file used in the previous exercise.

2. Move your cursor to the ViewCube and over the face labeled Front.

3. When the face highlights, click it to rotate your model.

4. Click the arrow that is pointing at the right side of the cube. This rotates your view to the right side.

5. Move your cursor to the upper-right corner of the ViewCube, and select it to see the top, right, and back faces of the ViewCube.

6. Click the ViewCube, hold your mouse button down, and move your mouse around to orbit the view.

7. After you've rotated your part, press the F6 key to return to the Home view of the assembly.

   Now, your screen should look as it did at the beginning of the exercise.

8. Click the face labeled Right.

9. After the view updates, right-click the ViewCube, hover over Set Current View As, and select Front from the menu that appears (see Figure 1.13).

FIGURE 1.13 Faces on the ViewCube can be redefined as Front or Top.

10. Orbit the view to something you think offers a better view than the current Home view.

11. Right-click the ViewCube and use Set Current View As Home with the Fit To View option.

The other option for defining a new Home view is Fixed Distance. This option maintains the zoom scale and position, even if additions to the model go outside the Graphics window.

The ability to define the orientation of the ViewCube and the Home view will prove useful as you gain experience.

Checking Out the Status Bar

On the bottom edge of the Inventor window is a bar that gives you information about expected inputs for tools. The status bar also gives you feedback on information needed by Inventor, as well as the number of files that are being accessed and parts that are being displayed when you are working in an assembly.

Autodesk 360 Sign-In

Autodesk 360 is a combination of many things. One thing it offers is online storage, which can be used to keep your files online and able to be accessed from any location. Autodesk 360 offers additional services you might like to explore. Signing in through the Taskbar will keep you connected.

Using Marking Menus

As you create drawing views, parts, and assemblies and as you use other tools, you will see dialog boxes for these tools.

Beginning with Inventor 2011, several "Heads-Up" interface tools have been added that put common commands at the mouse cursor. Marking menus work in two modes: Menu and Marking.

Menu Mode   When you are editing a file and right-click in the Graphics window, a "compass" (see Figure 1.14) of tools appears. As you move toward a tool, a shadow will highlight it, and clicking the mouse with this highlighting active selects the tool.

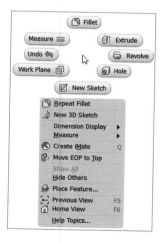

FIGURE 1.14 Marking menus place common tools where they are needed most.

In the Options panel of the Tools tab is a Customize tool. With it, you can change what tools appear on the compass and how they look. You can also control the size of the Overflow menu that appears beneath the compass.

Certification
Objective

Marking Mode   Once you're experienced with the tools offered in the marking menu, you can begin using Marking mode. To do so, right-click and drag the cursor toward the tool you want to select. When you release the button, you will activate the tool. The tool will flash to confirm that you've selected it rather than showing all the options.

Setting Application Options

Many settings allow you to tailor Inventor to your needs. Changing the tools in the Ribbon is just the beginning. To affect the way Inventor behaves, you will need to make changes to the application options.

To access the Application Options dialog box, open the Application menu and click the Options button at the bottom, or go to the Tools tab and select the Application Options tool. This will open a large dialog box with several tabs and an enormous number of options. I will focus on a few of these options, but I strongly recommend that you explore the others.

Once you are in the Application Options dialog box, you will see that the options are categorized in several tabs. I will suggest a few key settings for some of those tabs in the following sections.

Using the Import/Export Buttons

At the bottom of the Application Options dialog box are buttons that allow you to share your settings with others or export them for backup. You can also import one of these backups or import the settings another user has. Additionally you can change settings back and forth between those an experienced AutoCAD or Inventor user would be accustomed to.

Exploring the General Tab

Once you're comfortable with how you typically use Inventor, you might want to make changes to some of the basic settings. On the General tab, you can change how Inventor starts up, how long you have to hover over an icon before the tooltip will appear, and even how much room on your hard disk can be used by the Undo tool.

One option on this tab is to automatically update the physical properties of the model when you save. Enabling this option will keep things like the annotations added to the title block up-to-date without an extra step by the user.

Exploring the Colors Tab

The Colors tab has the favorite tools for exploration for most Inventor users. You can change the color and look of your Graphics window background, control the color of the sketch objects and dimensions, and change the color scheme of the Ribbon icons.

Follow these steps to save your current settings and experiment with some changes:

1. Keep using or reopen the c01-01.ipt file used in previous exercises.

2. Expand the Application menu, and click the Options icon.

3. At the bottom of the dialog box, click the Export button.
   A dialog box appears showing a couple of default settings backups. You can save your settings in this folder after you've made some changes.

   Export...

4. Click Cancel.

5. Select the Colors tab.

6. Select Presentation from the Color Scheme list, and notice the change in the preview window.

7. Drag the title bar of the dialog box to the left so you can see part of the Graphics window behind.

8. Click the Apply button at the bottom of the dialog box. See Figure 1.15 for reference.

9. Using the Background drop-down list, change the setting to 1 Color.

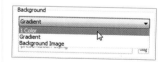

10. Click Apply to see the effect.

Now try experimenting with different Color Scheme values, as well as changing the color theme and background, until you find something that is pleasing to you. You can also use an image file for the background of your screen to give additional realism or create a special effect.

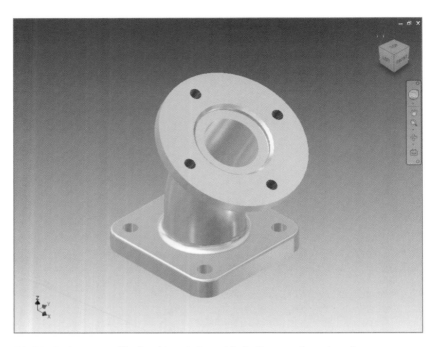

FIGURE 1.15 The Graphics window with the Presentation color scheme

For this book, I will use the Sky Color scheme with a white background image to keep the components in the Graphics window easy to see for the printed graphics. Using the Presentation scheme with a 1 Color background is a quick way to generate images for your technical documents as well.

Exploring the Display Tab

The Display tab controls a number of display effects you've already encountered. On the Display tab, you can use the Inactive Component Appearance section to control the fading effect on inactive parts while editing a part in the assembly.

Similarly, when you select the Home view, Inventor uses an animated transition to move the model into position. You can control the speed of that transition with the slider bar under Display.

Changing the Minimum Frame Rate (Hz) value to a higher number causes Inventor to make parts of an assembly invisible while zooming, panning, or orbiting in order to maintain the chosen frame rate. The real use of this value is to allow you to change the view quickly while still being able to see what position the model is in.

▶

When you are editing assemblies, enabling a shaded appearance and turning on the Display Edges option for Inactive Component Appearance creates an environment where the active component is the only one that is shaded. Using a subtle color for the display edges enhances the effect.

Other settings allow you to reposition the ViewCube on the screen and control the visibility of the Origin 3D indicator or its axis labels.

Exploring the Hardware Tab

Most of the settings for hardware are fairly self-explanatory. A very important item is the Software Graphics check box. If you experience frequent crashing in Inventor, you should select this check box (see Figure 1.16), restart, and see whether the crashing stops.

FIGURE 1.16 Selecting the Software Graphics option can help overcome stability issues.

The majority of crashes in Inventor begin as a conflict between the graphics driver, Windows, and Inventor. Selecting Software Graphics disconnects Inventor from communicating directly with the graphics driver. If the crashing stops, you should verify that your video card is supported and make sure you have the certified driver for the card. The Additional Resources section under Help has a link to look up the certified driver for approved video cards.

The Diagnostics tool at the bottom can help test whether the graphic drivers are Windows Hardware Quality Labs–certified and whether your graphics card is capable of offering acceleration.

Exploring the Assembly Tab

The Assembly tab's settings are primarily at-large assembly needs and specialized options. You can control whether the inactive components of the assembly remain opaque when a component is activated for editing; you can also change this option on the View tab by extending the Appearance panel.

The Express Mode Settings allow you to disable or set the file threshold for the number of parts in assembly that would trigger Express Mode. This mode limits what data is initially loaded when you open an assembly to only the graphics. This speeds up performance if you're not editing the assembly. If you want to edit, Inventor will need to load the background data.

If you decide that you would rather not hear the sound played when you place an assembly constraint, you can turn off that sound on this tab.

Exploring the Drawing Tab

The Drawing tab has some settings that should be very interesting to AutoCAD users. The Default Drawing File Type allows you to create drawings in the traditional Inventor IDW format, or you can choose to create DWG files that can be opened for viewing, printing, and measurement in AutoCAD. You can even set which AutoCAD release format a DWG will be created in.

You can set some overall dimension preferences, and you can control whether line weights are displayed. For users with limited graphics power, you can change the View Preview Display setting to see views with a full preview, with a partial view, or with a bounding box.

If you choose to create drawings in DWG format, they can be created in 2000, 2004, 2007, 2010, or 2013 format.

Exploring the Sketch Tab

In Chapter 3, "Introducing Part Modeling," you will begin exploring part modeling. When you first begin to use sketching tools, a grid can be displayed. For clarity, the graphics in this book will not display the grid, minor grid lines, or axes of the sketch. If you prefer not to see these, you can disable parts or all of the grid in the Display section of the Sketch tab.

In the Display settings, you can control how dimensions are displayed on entities as you sketch them. Many people prefer to disable the tools that automatically project edges during sketch or curve creation. These tools typically include Autoproject in the description. Many users also choose not to look at the sketch plane when sketches are created. You can adjust for those preferences here.

Exploring the Part Tab

A common change made on the Part tab is to enable a sketch creation when new parts are created. You can also choose to automatically have sketches created on a specific plane when you start a new part.

If you would like to begin new parts by sketching right away, you can set the plane you would like to begin on in the Sketch On New Part Creation frame.

Using Visualization Tools

The ability to communicate a design visually is a great asset of Inventor. Changing the color of the Graphics window or setting the background to be an image can give an interesting effect.

Increasingly, designers use renderings to present new concepts rather than creating detail drawings that might not be understood by people setting budgets. Regardless of the reason, the visualization capabilities that reside within the modeling environment can be just plain fun.

Understanding the Visual Styles

You can find the tools covered in this section on the View tab in the Appearance panel. I will review many of them, but I'm sure you will find yourself experimenting with them for a long time to come.

Certification Objective

Realistic   The Realistic visual style works at two levels. First, it uses the common color library that is shared among most Autodesk design products, giving you a different appearance than the other modes. Realistic also lets you very quickly create a raytraced rendering that is more than suitable for communicating a design concept. Although this option is in the modeling environment, it really isn't intended to be the setting you use while doing modeling or assembly functions. Not all graphics cards are capable of working with the raytracing capabilities.

Shaded   Shaded mode has been available since the first version of Inventor and is the environment that most users work in.

Shaded With Hidden Edges   To highlight the visible edges of the components and generate hidden edges that show through the components to make it easier to find faces that are otherwise obscured, use Shaded With Hidden Edges.

See Figure 1.17 for an illustration of these three modes.

Wireframe With Hidden Edges   Working in wireframe is comfortable for users who have worked with older 3D systems. The addition of displaying the hidden edges makes it easier to sort out what is in the foreground and the background.

Monochrome   Monochrome makes your model look like Shaded mode but with grayscale shades replacing the colors of the components. It is a useful tool to see how a noncolor print of your shaded model might look.

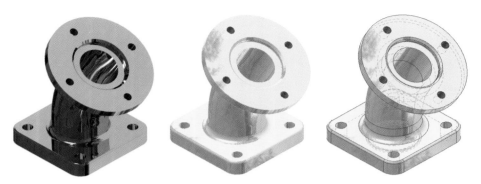

FIGURE 1.17   The Realistic, Shaded, and Shaded With Hidden Edges visual styles

Illustration   Illustration mode is in some ways a mirror to Realistic. It makes the model appear more like a line drawing by adding edges and varying line weights to the display.

See Figure 1.18 for the last three modes described.

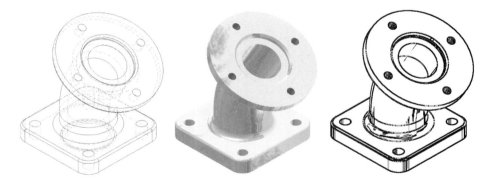

FIGURE 1.18   Wireframe With Hidden Edges, Monochrome, and Illustration visual styles

Using Shadows

Shadows can assist in highlighting the contours of a part, in particular when a part is prismatic and faces can be obscured in Shaded mode. These shadows can be displayed all at once, in any combination, or individually.

Ground Shadows   An easy way to add a visual appeal to your model is to have it cast a shadow on the ground. The position of the ground is controlled by the location of the top face on the ViewCube. The effect is of light shining down on the top of the model.

Object Shadows   Selecting the Object Shadows option causes components and features to cast shadows on one another based on the direction of the lighting.

Ambient Shadows   Ambient Shadows mode gives a somewhat more subtle but very effective change to the model but adds the effect of ambient occlusion, which causes areas where sharp edges meet to be darkened. See Figure 1.19.

FIGURE 1.19   With the model in Shaded mode, turning on Ambient Shadows gives it a little more realism.

At the end of the Shadows flyout menu is a Setting button. When you click that button, Inventor opens a dialog box where you can establish the direction from which the shadows will be cast. You can also control the density and softness settings, as well as the intensity of ambient shadows. You can also turn on image-based lighting here; you will look at that effect a little later in the chapter.

Using Ground Reflections

 Ground reflections are based on a calculated "ground" that lies beneath the lowest component feature or visible sketch point. The ground establishes its location based on the position of the top face of the ViewCube. The Ground Reflections option is either on or off, but the fun doesn't have to stop there.

Like the Shadows tool, Ground Reflections provides settings in a flyout menu next to the icon. In the settings, you can control the level of reflection, how much it is blurred, and how quickly that blur falls off. This dialog box also controls the settings for the Ground Plane tool, which will be described next.

Using the Ground Plane

 This option is just a switch that controls whether a grid will be displayed on the ground that the reflections and shadows are cast on. As mentioned, its settings are shared with Ground Reflections but can be accessed from its own flyout menu.

Putting Visual Styles to Work

In this exercise, you will get a chance to explore a couple of the options just reviewed:

1. Keep using or reopen the c01-01.ipt file used in previous exercises.

2. Switch your Ribbon to the View tab.

3. In the Appearance panel, click the Shadows icon and observe the change to the screen.

4. Right below the Shadows tool, click the Ground Reflections icon.

5. Locate the Ground Plane tool, and turn it on.
   You will now use some of the tools that were discussed earlier.

6. Use the Visual Style drop-down menu to set your style to Realistic.

7. Change your Visual Style setting to Watercolor. Note that the reflections and ground shadows disappear but ambient shadows are still generated.

8. Set Visual Style to Shaded when you're done admiring the Watercolor style, and keep this file open.

With a nice basic view, let's make a couple of simple changes to see what the effects are before looking at a few more visualization options:

1. Open the Shadows settings dialog box by using the flyout menu next to the Shadows icon.

   If possible, reposition the dialog box to see the screen as you make the following changes (because the screen will update while you are making them).

2. At the bottom of the dialog box, use the drop-down menu to change Shadow Direction to Light1.

3. Use the slider bar below to change the Density value to 40.

4. Change the Softness value to 80.

5. Reduce Ambient Shadows to 40.

6. Click the Save button, and then click Done to close the dialog box. See Figure 1.20, and keep the file open for the next exercise.

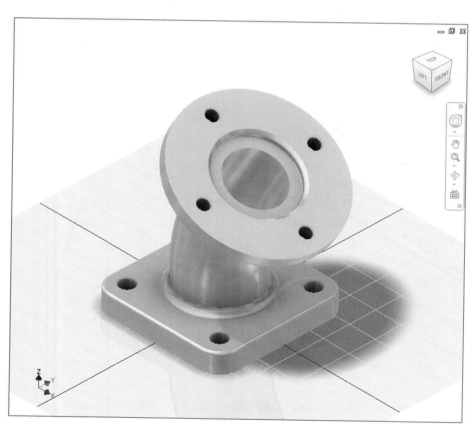

FIGURE 1 . 2 0 Small changes can create a dramatic effect with visualization settings.

With your shadows updated, you can do a little experiment with the reflections:

1. Open the Ground Reflections settings dialog box using the flyout menu next to the Ground Reflections icon.

   If possible, reposition the dialog box to see the screen as you make the following changes (because the screen will update while you are making them).

2. Set Reflection Level to **50%**.

3. Set Blur Level to **50%**.

4. Change the value in the Minor text box under Appearance to **1**.

5. Remove the check mark from Plane Color.

6. Set the Height Offset value at the top of the dialog box to **13mm**.

7. Click OK to save the changes and close the dialog box.

8. In the Appearance panel, find the drop-down menu for orthographic and perspective projections, and set the view to Perspective.

> Another option, Perspective With Ortho Faces, can be turned on by right-clicking the ViewCube and selecting it.

9. Press the F6 key to restore the Home view, and compare your screen to Figure 1.21. Keep the file open for the next exercise.

Remember, these changes aren't just for a printed document; you're still in a working part file, and you are able to make edits with these view characteristics.

Setting the Lighting Style

Lighting Style is an understated name for a setting. Lighting styles do, in fact, change how lights (and therefore shadows) are positioned in your model, but several of them also come with an environment that offers the effect of putting the model into a different world.

Through these settings, you can make the same changes to the lighting style that you were able to make to shadows, but you will be limited in the built-in selections of image lighting sources. This tool's effect is easier seen than described, so let's use it.

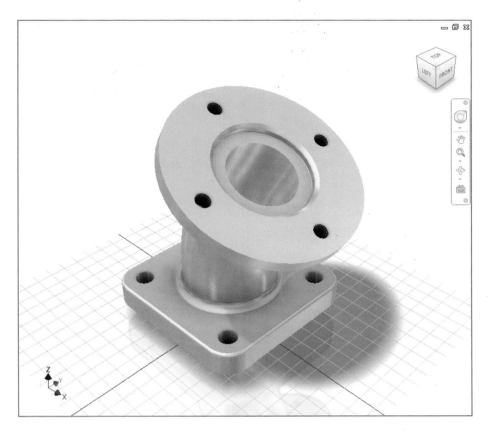

F I G U R E   1 . 2 1   Some changes might decrease realism but can add energy to the view.

You will do only a couple of changes, or you may not get any further in the book. Some of these next changes might not work with all graphics cards.

1. Continue working in the open file.

2. In the Appearance panel, the Lighting Style drop-down currently shows the Two Lights style. Change its value to Old Warehouse.

3. Use the ViewCube or Orbit tool to rotate your part around and see the environment that is shown. Using Pan and Zoom can help with finding some interesting points of view too.

4. Change your lighting style to Empty Lab.

Setting a lighting style can change a lot of things about how you look at and work with your model. It's also a lot of fun.

Using Color Override

When you begin to create 3D part models, you will use specific materials so that you can monitor and predict the weight and inertial properties of the components you are building. Materials have their colors assigned to them, but you are not limited to having steel look like steel, for example. Using the Color Override option, you can change its color to make it look like anything you want.

Inventor 2014 offers more than one library of colors and materials. The Inventor Material library includes colors that have been used for years by Inventor. There is also an Autodesk Appearance Library and an Autodesk Material Library. Using colors from the Autodesk libraries give you consistency in the appearance of the parts if you open them in 3ds Max® or other Autodesk applications.

Changing a component's color is straightforward, but keep in mind that changing a component's color in the assembly does not change it in the source file. If you want to change the color of something everywhere, you need to do it in the part file.

How easy is it? Let's find out:

> If the color you select has a texture, it will make the faces of the part look like there are additional features or real textures on them. To maintain graphics performance, you can switch the display of textures on and off in the Appearance panel of the View tab.

1. Continue working in the open file.

2. Find the Color Override drop-down in the Quick Access toolbar and select a different color.

   At the bottom of the Color Override drop-down you can change the library used for color selection.

3. Change the library to the Autodesk Appearance Library, and set the color to Chrome – Polished Blue.

4. As a nice finishing touch, change your visual style to Realistic. See my results in Figure 1.22.

You can close the c01-01.ipt file by clicking the X in the upper-right corner of the Graphics window or through the Application menu. (There is no need to save the results of this exercise.)

One of the fundamental steps in learning the Inventor tools is to know where your files are. For that, you need a project file. The project file can also predetermine several options for accessing libraries, including the appearance and materials libraries. Next you will see how to create one.

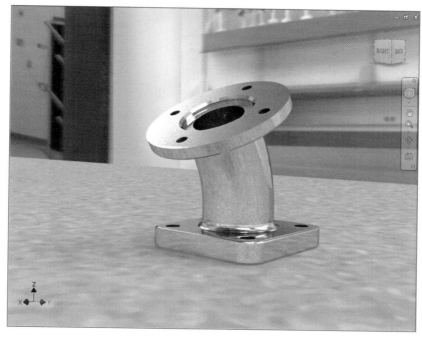

FIGURE 1.22 There is nothing some form of Chrome can't improve.

Working with Project Files

To work with Inventor effectively, you must have an understanding of the role of the *project file*. The project file is a configuration file that tells Inventor where to look for and store data about a project. Among other things, the project file can be used to create shortcuts for locating files, change where Inventor looks for templates, and control the user's ability to modify standards.

It's also important if you work with others to have a consensus on how you will work with data and consistency in your project files. Working from the same project file can save a lot of frustration. This can easily be done by simply copying the file among several users.

By now, you should've extracted the data for the files you will use in the book. Now, you will tell Inventor where to find that data.

Certification Objective

Creating a Project File

You can create a project file at any time, but you can make it active only when all data is closed in Inventor. Now you will create the project file to access the data for this book.

1. Make sure that all files are closed in Inventor but keep Inventor running.

2. On the Get Started tab, click the Projects icon in the Launch panel.
   This opens the Projects dialog box, where you can control the active project file or edit existing ones.

3. Click the New button at the bottom of the dialog box.
   This launches the Inventor Project Wizard, which will walk you through the process the next time you want to create a new project file.

4. Click the New Single User Project option, and then click the Next button at the bottom of the dialog box.

5. Enter 2014 Essentials in the Name field.

6. Enter `C:\Inventor 2014 Essentials\` in the Project (Workspace) Folder field.

7. Compare your settings to Figure 1.23, and when you're sure everything matches, click Finish to create your new project file.

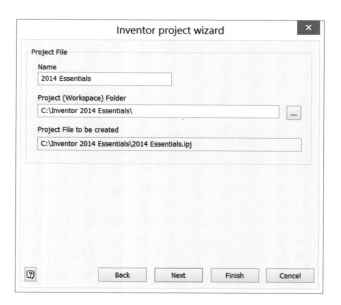

FIGURE 1.23 Defining the name and location of the new project file

Now you have a 2014 Essentials.ipj file in the folder that you specified. (The .ipj extension stands for Inventor Project File.) You need to add some detail to make sure everything will be controlled as needed.

Modifying the Project File

In this exercise, you will add shortcut paths to folders that you will frequently access and make sure that the standard parts you use are kept where you need them.

1. In the top window of the Projects dialog box, double-click the 2014 Essentials project file to make it active.

2. In the bottom window, right-click Use Style Library, and switch it to Read-Write in the context menu.

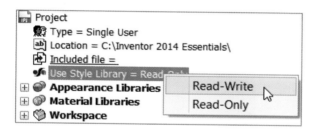

3. Click the Frequently Used Subfolders icon to highlight the row.

4. On the right, click the Add New Path icon.

5. A new row will be added under the row. In the field on the left, type **Assemblies**. On the right, browse to set your path to C:\Inventor 2014 Essentials\Assemblies. See Figure 1.24.

6. Click OK to set the path for the shortcut to the assembly files.
   When the new shortcut is created, you will see that the path is different from what it was in the dialog box. Instead of showing the full name, it will say *Workspace\Assemblies*. This helps avoid problems if the files need to be moved to another drive.

7. Add a shortcut for parts pointing toward the Parts folder and for drawings pointing to the Drawings folder.

8. Expand the Folder Options section.

9. Click the Content Center Files icon, and click the Edit icon on the right.

10. Set the path for the Content Center files to C:\Inventor 2014\ Content Center Files, as shown in Figure 1.25, and then click OK to complete the edit. When prompted, save the changes.

As mentioned, the project file is critical for working with other users on the same data. The Content Center files path can be set to a common network location so that everyone can find the correct information when editing each other's work.

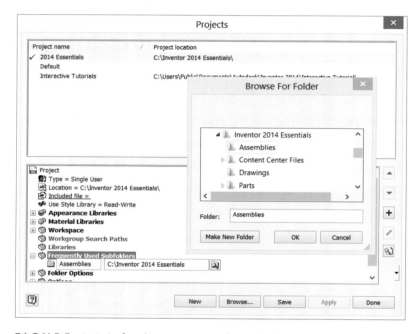

FIGURE 1.24   Creating an easy-to-understand alias representing a file path

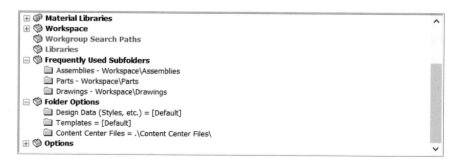

FIGURE 1.25   Setting the path for saving standard components used in an assembly

THE ESSENTIALS AND BEYOND

As you get more experience with the Inventor interface, you will see that it assists you in understanding which environment you are in and what is available to you. Changing colors and setting options that you like is a great way to get more comfortable, and that will lead to you being more productive.

ADDITIONAL EXERCISES

▶ Add the tools that you think you will find useful to your Quick Access toolbar.

▶ Challenge yourself to access tools from different locations to become familiar with other tools around them.

▶ Use different application options, especially for sketching and part modeling, to see the effect they have on your productivity.

▶ Experiment with color schemes and lighting styles to find ways of maximizing the impression you make on people who see your work.

CHAPTER 2

Introducing Parametric Sketching

When creating 3D geometry using the Autodesk® Inventor® 2014 software, you must begin with a sketched feature. To have a sketched feature, you must have a 2D sketch. In this chapter you'll use a majority of the tools that are available for creating the 2D sketches so critical to creating your designs in Inventor. You'll also learn how to control this geometry beyond the basics, and we'll briefly touch on 3D sketching, which can be used to extend the capabilities of 2D sketches.

▶ **Exploring the essential elements of a 2D sketch**

▶ **Controlling a sketch**

▶ **Using dimensional sketch constraints**

▶ **Defining and placing sketches**

Exploring the Essential Elements of a 2D Sketch

The 2D parametric sketch that you create in Inventor has several components. The most familiar to users of 2D CAD systems, or even to people who've only used pencil and paper, are the geometric elements of lines, arcs, and circles. In a parametric sketch environment, these basic pieces are enhanced with constraints, dimensions, and special formatting options—which means you no longer have to erase and redraw if a feature of your design changes size.

Before you start sketching, let's look at an example of how constraints work.

Controlling a Sketch

Certification
Objective Rather than beginning with creating simple geometry, you should know what is happening as that geometry is created to give it its power. To do so, you will add geometric constraints to a sketch that was deliberately created without the common constraints that are automatically created when you begin sketching.

Coincident Constraint

For you to create a 3D feature from a sketch with more than one element, there must be *coincidence* between the endpoints of those elements. The coincident constraint will normally be created by default. In this example, you will see how to add it if geometry is created in a disjointed fashion.

1. Verify that the 2014 Essentials project file is active, and then open the c02-01.ipt file from the Parts\Chapter 02 folder.

    In the Design window you will see four line segments that you will convert into a rectangle (which you will be able to control the size of) and a point that represents the intersection of the part's x-, y-, and z-axes. To understand how to constrain these elements, you first need to know how or if there are constraints on the geometry you already have.

2. Locate Sketch1 in the browser and double-click or right-click it. Then select Edit Sketch to edit the existing sketch.

    Once the sketch is active, the status bar will update and a message at the bottom of the screen will inform you that you will need 16 dimensions to have complete control of this sketch. In this case, *dimension* is a generic term for a control. As you apply constraints to the geometry, the number of dimensions required will be reduced as well.

3. Right-click in an open area of the Design window and select Show All Degrees of Freedom from the context menu to display special icons that represent the ability of the lines and their endpoints to move or rotate, as shown in Figure 2.1.

4. To begin connecting the endpoints, start the Coincident constraint from the Constrain panel of the Sketch tab.

5. Pick the endpoints of two lines where they nearly meet to connect them into a corner.

6. Repeat step 5 for two more corners; then press Esc to end the Coincident tool.

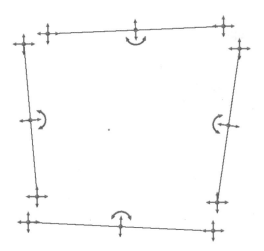

FIGURE 2.1 All four line segments can move
or rotate on the sketch plane.

7. For the final corner, click and drag one free endpoint to the other to
   form the fourth connected corner. As you near making the connection the coincident points will highlight in green and a glyph will
   appear showing a point moving onto another point.

Your sketch should resemble Figure 2.2. The sketch will now only
need eight dimensions. Your sketch is now a "closed loop," which
means that all of its members are connected and form an intact profile. This is all the information Inventor needs to create a 3D feature.

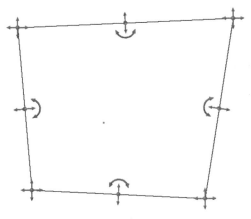

FIGURE 2.2 Applying coincidence makes the
sketch a closed loop.

8. Click and drag various edges and points at the corners to see how, while flexible, the sketch stays connected. Return the sketch to a state similar to Figure 2.2, with the projected center point in the middle of the quadrangle.

Common Constraint Options

With a basic understanding of the power of a simple Coincident constraint, you will now work through some of the other constraint options:

1. Continue using the c02-01.ipt part with Sketch1 active.

2. Locate the Perpendicular constraint tool in the Constrain panel of the Sketch tab.

3. Select the line on the left and then the bottom line to force them to be perpendicular to one another.

4. Start the Parallel tool, which is also on the Sketch tab and in the Constrain panel.

5. Pick the line on the left and then the line on the right to see how they align to each other.

6. Start the Horizontal constraint tool and select the top line. Notice that the curved arrow disappears. This is because the Horizontal constraint removes the line's ability to rotate.

7. To remove all of the rotational degrees of freedom in the sketch, start the Vertical constraint tool and select the line on the left. This will leave the sketch looking like Figure 2.3.

Your sketch now needs only four dimensions. You can still move the elements in the X and Y direction but you cannot rotate them. Let's bring the sketch in line with the origin planes of the part.

Alternate Points for Alignment

You've constrained and even moved the endpoints of the sketch. Inventor also understands the value of being able to connect to the midpoints of lines and arcs. These midpoints will highlight in green just like the endpoints did.

1. Start the Horizontal constraint and select the midpoint of the line on the left and then the origin center point to center the vertical lines along the y-axis of the part.

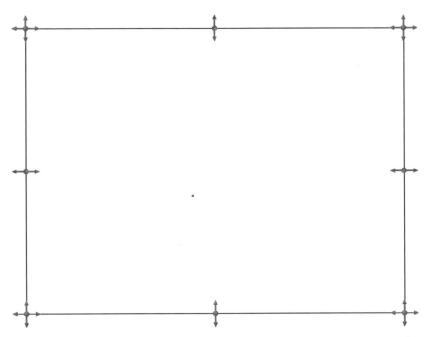

FIGURE 2.3 The lines are now barred from rotating thanks to a combination of constraints.

2. Use the Vertical constraint on the midpoint of the top line and the origin center point.

3. Select the Finish Sketch tool in the Sketch tab's Exit panel to end the constraint tool and exit the sketch.

The sketch now needs only two dimensions (see Figure 2.4). These dimensions will control the size of the rectangle along the X and Y direction of the part. Geometric constraints are powerful due to their persistence. Now that you've made the first two lines perpendicular, they will remain that way until you deliberately change that relationship. Dimensional constraints are useful for controlling geometry that might need to be changed more frequently.

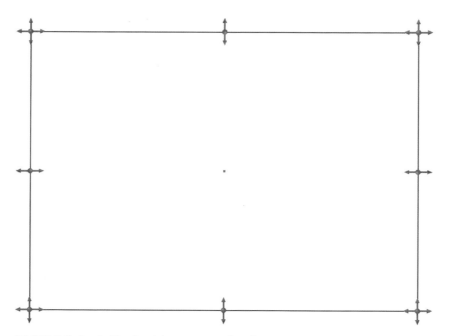

FIGURE 2.4 The sketch is now centered on the part's origin planes.

 NOTE Inventor does not require that you relate sketches to the origin point in any way. However, if you are making a symmetrical part centering your sketches on the default planes and axes makes sense.

Further Exploring Geometric Constraints

With an understanding of basic constraints, you can now use a variety of other constraints to clean up an incomplete sketch:

1. Verify that the 2014 Essentials project file is active, and then open the c02-02.ipt file from the Parts\Chapter 02 folder.

2. Edit Sketch1 using the browser or by selecting an element of the sketch.

3. Start the Fix constraint that is in the Constrain panel of the Sketch tab.

4. Pick the center point of the circle in the upper right of the sketch to lock its center in place.

5. Press Esc to end the Fix constraint.

   To use the circle as a focus of the sketch, you can refer other geometry to it, beginning with the arc. It is not necessary to exit one constraint or tool to start another. For the next series of steps, you can exit one tool by starting another.

6. Start the Concentric tool from the Constrain panel and select the edge of the circle and then the edge of the arc.

7. Use the Collinear tool from the Constrain panel to align the two short lines at the top of the sketch, as shown in Figure 2.5.

   To align the two lines to the center of the circle, use a Coincident constraint.

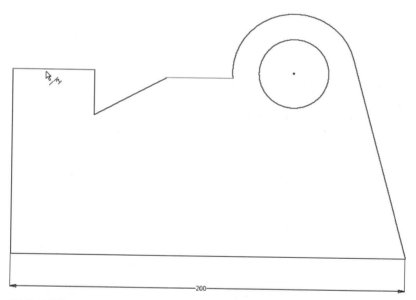

FIGURE 2.5 Selecting lines to align them to each other

8. Choose the Coincident tool from the Constrain panel and select the center point of the circle; then pick one of the two lines that you applied the Collinear constraint to. Be sure to pick the line and not an endpoint or center point of the line.

9. Make the short, angle line tangent to the top of the arc b using the Tangent constraint in the Constrain panel.

10. Select the Finish Sketch tool in the Exit panel of the Sketch tab.

Now you have finished applying geometric constraints to the sketch and it should look like Figure 2.6. Next you can add dimensions to completely control the sketch geometry.

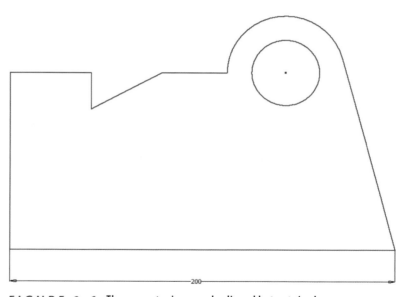

FIGURE 2.6 The geometry is properly aligned but not sized.

Using Dimensional Sketch Constraints

Certification
Objective

The skill of the drafter and the AutoCAD user is centered on accurately drawing geometry to a precise size and then documenting that precision with a dimension that can be used by manufacturing to re-create that geometry. When you work with parametric sketches the process may seem different, but in fact you are just using the dimension to control the precise creation of the geometry. The line won't be 350 mm long because of the drafter's skill; it will be that precise size because of the dimension's value.

Getting Started with Sketch Dimensions

This process is something that is easier done than said as you will see in the following steps:

1. Verify that the 2014 Essentials project file is active, and then open the c02-03.ipt file from the Parts\Chapter 02 folder.

2. Click any portion of the sketch and select Edit Sketch from the mini-toolbar or double-click Sketch1 in the browser.

   It is always a best practice to create your sketches as close to the correct size as possible. Inventor has a great tool for times that you don't quite get the size right. The first dimension you place on the sketch will scale the sketch to keep the various elements in the correct perspective.

3. In the Constrain panel of the Sketch tab, start the Dimension tool.

   Dimension

4. Click the bottom line. This will create a dimension preview that shows the current size of the line.

5. Click to place the dimension below the rectangle. When the Edit Dimension dialog box appears, set its value to **150 mm**, as shown in Figure 2.7, and click the green check mark to update the sketch.

   The sketch will update and appear to vanish because it is larger than the Design window.

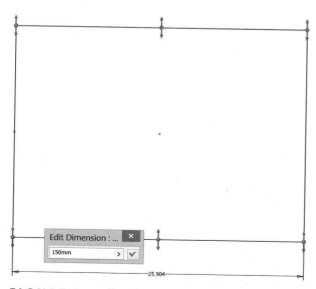

FIGURE 2.7 Changing the dimension will change the length of the line.

6. Click the View All icon in the Navigation bar or press the Home key to update your drawing view.

7. The Dimension tool will still be active, so click the vertical line on the left and place the dimension to the left of the sketch.

   As you place dimensions in the sketch, Inventor assigns a name to them behind the scenes so that it is easy for the system to build relationships. There are several ways to take advantage of this naming; the most basic is to relate one dimension to another.

8. When the Edit Dimension dialog box appears, it will display the current value. When this happens, pick the first dimension.

   The name of the first dimension, d0, will appear as the value of the new dimension. If you press Enter to accept this value, the height of the new dimension will be 150 mm.

   For this example you will set the height to be a factor of the width and, rather than using the automatically generated dimension name, you will specify one.

9. Use your mouse to put the editing cursor behind the d0 value, and then type **/2** to specify that you want this dimension to be half the value of the referenced dimension.

10. Now move your cursor before the d0 value and type **Height =** to specify a name for the new dimension rather than accepting the default value (d1); then click the check mark to accept the new value.

    Adding this dimension will also remove the last needed dimension. The status bar will now report that the sketch is fully constrained.

11. Press Esc to end the Dimension tool.

12. Double-click the 150 mm value, set the value in the Edit Dimension tool to **200**, and press Enter to accept the value.

13. Right-click Sketch1 in the browser and select Finish 2D Sketch from the context menu.

    The rectangle will get larger. Because the part was made using a metric template, there is no need to type the unit. If you want to use a different unit of measure (cm, m, in), you can specify it with the value.

Now that you've set up the size and shape of the sketch, it is time to see how a sketch can be changed once it is set up.

Changing the Constraint Structure of a Sketch

Understanding the process of editing an existing sketch is critical if you have to change someone else's work or if a sketch needs to change in a way that wasn't originally anticipated. Learning a few basic editing tools and techniques will make it possible for you to change any sketch:

1. Verify that the 2014 Essentials project file is active, and then open the c02-04.ipt file from the Parts\Chapter 02 folder.

    To edit this sketch, you need to see how it is structured beyond the obvious dimensions.

2. Edit the Sketch1 through the browser or by clicking the sketch.

3. Right-click and select Show All Constraints from the context menu, or press the F8 key.

    Using Show All Constraints turns on the icons that show what geometric constraints have been applied to the geometry. The constraints that have been applied are represented by icons that look like the icons for the constraint tools that applied them. To keep the screen less cluttered, coincidence constraints appear as yellow squares at the corners where you applied coincident constraints.

4. Find the Parallel constraint on the vertical line to the right. Move your cursor over the icon. As you do this, notice that the line on the left highlights, showing you what the parallel relationship is connecting to.

5. Right-click and click Delete in the marking menu. This will remove the constraint, and once again your sketch requires a dimension (or constraint) to be added in order to be fully controlled.

6. Do some experimentation with clicking and dragging the left and right edges of the rectangle to see how the line on the right will

rotate around its midpoint, which is still held in position by the constraints that keep the sketch centered.

7. Start the Dimension tool on the Sketch tab in the Constrain panel. Pick the bottom line of the sketch and then the right edge. As you near the right edge the glyph near the cursor will change to an angle dimension, indicating that Inventor will add an angle dimension to the sketch.

8. Click to place the dimension to the left, and set the value of the dimension to **60** to make your sketch look like Figure 2.8.

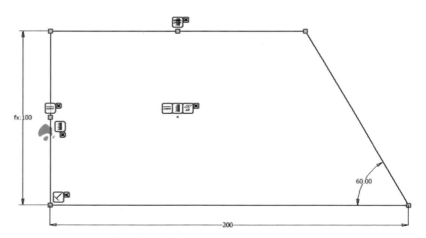

FIGURE 2.8 Changing the sketch can be as easy as removing a constraint and adding a dimension.

9. With the Dimension tool still active, pick the angled line. Move the cursor to the right or up to see the dimension preview switch from vertical to horizontal.

10. Right-click, select Aligned in the context menu, and pick where you want the dimension.

When you do this, a warning box appears telling you that the new dimension will overconstrain the sketch and offering you a chance to place a driven dimension. A driven dimension displays a value and is associative to the geometry, but you cannot set its value.

11. Click Accept to place the dimension and then press Esc to end the Dimension tool.

    If you want to control the height of the overall sketch using the length of the angled line, you can easily do so by changing which dimensions are driven and which are parametric.

12. Pick the vertical dimension on the left to select it and click the Driven Dimension Override tool in the Format panel of the Sketch tab.

    Doing so changes the vertical dimension to a reference and also reopens a degree of freedom in the sketch, thus allowing the height to be changed.

13. Pick the aligned dimension on the right to select it; right-click and choose Driven Dimension Override on the marking menu to convert the dimension to a parametric dimension.

14. Change the value of the aligned dimension to update the sketch so it matches Figure 2.9.

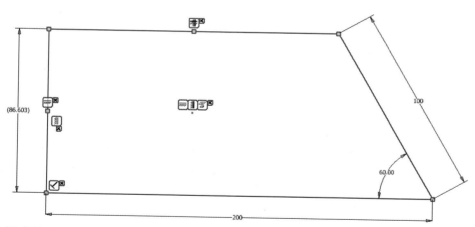

FIGURE 2.9 Changing the dimensions that control the sketch

15. To turn off the constraint visibility, right-click and select Hide All Constraints, or press the F9 key.

With just a few clicks, you can change the control hierarchy of a sketch. This flexibility is why parametric solid modeling has become such a dominant technology in mechanical design.

Defining and Placing Sketches

A sketch can be created on any plane in your part. You can use existing faces or you might need to create a plane to sketch on. In Chapter 6, "Exploring Part Modeling," you will have an opportunity to work with a number of different ways to add planes to your parts.

Creating a Sketch

There are various ways to create a sketch, but they all rely on an existing plane to align the sketch to. In the following steps you will use the three most common methods to create a sketch and use some new sketching tools along the way:

1. Verify that the 2014 Essentials project file is active, and then open the c02-05.ipt file from the Parts\Chapter 02 folder.

Create
2D Sketch

2. Click the Create 2D Sketch tool in the Sketch panel of the 3D Model tab.

    When you tell Inventor that you want to create a 2D sketch in a new part file, it will require you to select one of the part's origin planes to align the sketch to.

3. In the browser, expand the Origin folder and select the XY Plane on the screen, as shown in Figure 2.10.

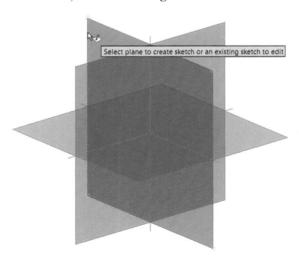

Select plane to create sketch or an existing sketch to edit

FIGURE 2.10 Selecting a plane to place a new 2D sketch on

4. Select the Rectangle tool in the Draw panel of the Sketch tab.

⬜ Rectangle

5. Click a spot in the Design window below and to the left of the part origin, and drag the rectangle up and to the right.

    As you drag, you will see a value for the height and width of the rectangle. Simply picking a second point would create a rectangle with no dimensions, which is great for creating a rough sketch.

6. The Horizontal Dimension value is highlighted as you are dragging. Type the number **25** and it will replace the value in the dimension and prevent you from dragging the rectangle until you lock in the value.

7. Press the Tab key to set the Horizontal Dimension value and enable you to enter a value in the Vertical Dimension field.

    Once you set a horizontal dimension, you can drag the rectangle vertically but its width is set.

8. Type **15** and press Enter to create the dimensioned rectangle in Figure 2.11.

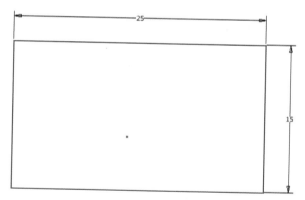

FIGURE 2.11 Tools like Rectangle make sketching easy.

9. Right-click and choose Finish 2D Sketch from the marking menu.

A new sketch has been added to the browser. When creating 3D models you may create many sketches in a part, and as you add 3D features to those sketches, they will be "consumed" by those features. Let's add sketches to a component with existing features:

⊗ End of Part

1. In the browser, click and drag the End Of Part icon to the bottom of the browser and release it.

2. Press the F6 key to bring the model to the middle of the Design window.

3. Select the small rectangular face on the left face of the part as shown in Figure 2.12 and select Create Sketch from the mini-toolbar that appears near the cursor.

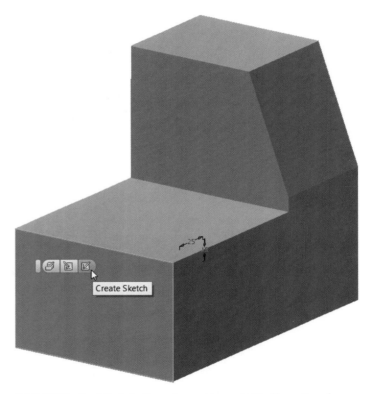

FIGURE 2.12 Selecting a face on the part will offer options for editing the selected feature or creating a new sketch.

By default the part will rotate to look directly at the new sketch.

⊘ Circle ▾

4. Select the Circle tool from the Draw panel of the Sketch tab.

5. Move the cursor near the center of the face that you selected to find the projected point aligned with the center of the part. Click and drag the diameter of the circle; click a second time to create the

circle. Do not enter a diameter value so you can see that the circle can be created without a specific size.

6. Right-click and select Finish 2D Sketch to complete the sketch.

7. Press F6 to restore the Home view of the part.

Creating a Sketch on a Work Plane

At times a sketch may need to be created a distance off an existing face. This sketch will need to be created on a special feature called a work plane. You will learn more about work planes in Chapter 3, "Introducing Part Modeling," but for now you will see how to create a work plane at the same time you create a new sketch.

Certification Objective

1. Start the Create 2D Sketch tool again; this time click and drag the preview away from the angled face 50 mm (see Figure 2.13). Click OK on the mini-toolbar to create a work plane and a new sketch on that plane.

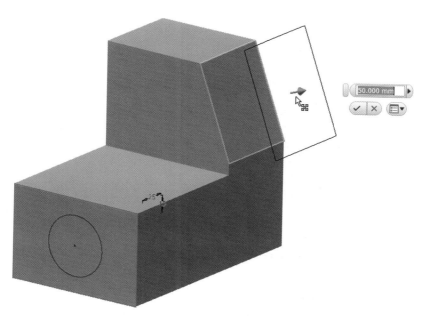

FIGURE 2.13 An offset work plane can be created along with a new sketch.

2. Start the Ellipse tool from the Draw panel on the Sketch tab.

⬭ Ellipse

3. Click near the center of the face you offset the new sketch from to set the center of the ellipse.

4. Drag your cursor up and you will see a glyph appear, indicating when the first axis of the ellipse is vertical. Go up a short distance and click to set the size of the first axis.

5. With the first axis defined, you will see a preview as you click to place a second, shorter axis; then right-click and select Cancel (Esc) from the marking menu.

   The precise size and location isn't important because you will dimension the location and size of the ellipse using parametric dimensions.

6. Start the Dimension tool from the Sketch tab's Constrain panel.

   When you dimension an ellipse, you dimension half of each axis length.

7. Click the ellipse and place a vertical dimension, set its value to **1.5 in** to see that units are flexible, and then set the horizontal dimension to **18 mm**.

8. Add a dimension by clicking the ellipse and then clicking the top edge of the angled face to set the location of the ellipse center to 45 mm from the edge.

9. Add a horizontal dimension between the center of the ellipse and the far left edge of the part; set its value to **165 mm**. Refer to Figure 2.14.

10. Press Esc to close the Dimension tool.

11. Right-click and select Finish 2D Sketch from the marking menu.

    Inventor will automatically project an edge into the sketch to connect the last dimension to.

There are many ways to create a work plane. Work planes do have other functions, but establishing a place to put a sketch is among the most common purposes.

Exploring the Dimension Tool

The Dimension tool in Inventor can place just about any type of dimension in the sketch with little or no special treatment. If you're an AutoCAD user, you might expect to have to use a lot of different tools to do the same work that Inventor can do with one tool. Its ability to understand what type of dimension is desirable keeps you focused on constraining the sketch rather than picking different types of dimension.

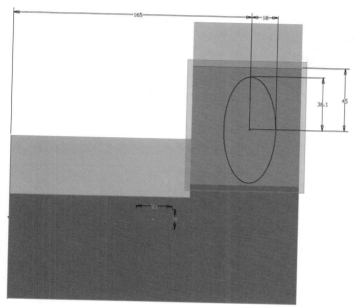

FIGURE 2.14 Dimensions can be placed based on edges that are not aligned to the sketch.

In the following steps you will detail an otherwise constrained sketch to be dimensioned like Figure 2.15. Please refer back to Figure 2.15 to see how the completed model should look.

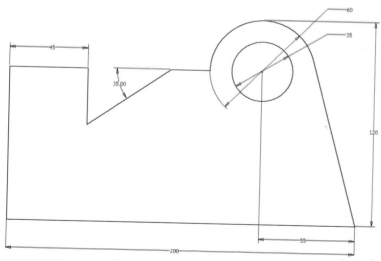

FIGURE 2.15 The fully dimensioned sketch

1. Verify that the 2014 Essentials project file is active, and then open the c02-06.ipt file from the Parts\Chapter 02 folder.

2. Edit Sketch1 using the browser or by selecting an element of the sketch.

3. In the Constrain panel of the Sketch tab, start the Dimension tool.

4. Select the edge of the circle; as you near the circle, a glyph will appear next to the cursor to show that it will create a diameter dimension.

5. Pick the edge of the circle, place the dimension, set the value to **35**, and press Enter or click the check mark to set the value.

6. Without exiting the Dimension tool, move to the arc, which will change the glyph to a radius; select it and then right-click to open the context menu.

7. In the context menu, expand the Dimension Type submenu and select Diameter.

8. Place the dimension, which will be a diameter, set the value to **60**, and press Enter to accept the value.

    You can also change a radius dimension to an arc length and a diameter dimension can be set to be a radius.

9. Make the horizontal line in the upper left **45 mm** wide.

10. Add an angular dimension between the short angled line and the horizontal line to its right, placing a 35-degree dimension in the "notch" of the part.

    The next dimensions are linear dimensions. Be careful about what geometry you select. Linear dimensions can be placed by selecting a single line, two parallel lines, a line and point, or two points. A linear dimension can also be placed between a line and the extents of an arc through careful selection.

11. Click the bottom line of the sketch first; then move your cursor to the top of the arc. You may need to move back and forth along the top of the arc until a glyph appears showing a dimension to the tangency of the circle.

12. Select the arc and set the overall height of the part to **120**.

13. Now pick the center point of the circle and the endpoint of the line that is at the lower-right corner of the sketch.

14. Use these selected points to create a horizontal dimension of **55** to complete the sketch. Press Esc to close the Dimension tool.

15. Finish the 2D sketch; you can save the file if you like.

The dimensions and constraints make it easy to control the size and behavior with simple edits to the assembly constraints. There are additional capabilities for modifying and controlling dimensions covered in Bonus Chapter 3, "Automating the Design Process and Table-Driven Design."

Going beyond Basic Lines

Most of the sketching and constraint tools of Inventor are simple in their operation. Where there are techniques or shortcuts for getting more out of a tool, the process for getting more is simple as well.

1. Verify that the 2014 Essentials project file is active, and then open the c02-07.ipt file from the Parts\Chapter 02 folder.

2. Edit Sketch1 using the browser or by selecting an element of the sketch.

3. Start the Line tool from the Draw panel of the Sketch tab.

4. Move near the left side of the larger circle. When the yellow dot of the cursor attaches to the edge of the circle, click and hold the mouse button and drag down and to the right. As you do, you should see a glyph that shows that the new line is tangent to the circle.

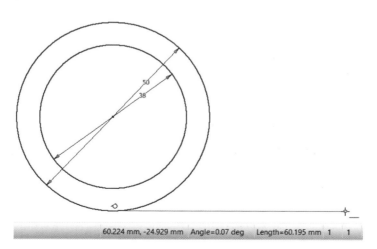

60.224 mm, -24.929 mm   Angle=0.07 deg   Length=60.195 mm   1   1

FIGURE 2.16 Dragging a line off a circle or arc will create a tangent line.

5. Drag to be horizontal using the glyph (see Figure 2.16), and when the status bar show the line is roughly 60 mm long, release the mouse to create the line.

6. Move your cursor back to the endpoint of the line. When it highlights with a gray point, click and drag the mouse to the right and it will begin creating an arc tangent to the line.

7. Drag the tangent arc up. Notice that a dotted inference line appears as you bring the endpoint directly above the origin to make a 180-degree arc (see Figure 2.17).

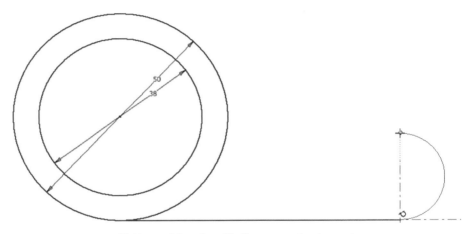

FIGURE 2.17 Clicking and dragging off a line can create a tangent arc.

8. Release the mouse button to create the arc.

9. Drag the line toward the outer circle. When the line is showing that it is coincident to the circle and tangent to the arc, click to create it.

10. End the Line tool by pressing Esc or by choosing Cancel from the marking menu.

    If you try to drag a tangent line off an arc and it goes the wrong way, you can quickly move the cursor back past the first point to reverse direction. Similarly, you can drag an arc normal to a line by moving it normal to the line's axis rather than along it.

11. Select the Center To Center Slot tool in the Draw panel of the Sketch tab.

12. Pick the center point of the tangent arc as the start point.

13. Move the cursor to the left, keeping the slot horizontal, and set the length value to 35. Press Enter to set the diameter.

14. Move your cursor to set the radius of the slot. Click to keep it smaller than the outer sketch but don't set a value.

15. Start the Dimension tool and click the right radius of the slot. Then pick the outside radius concentric to it to create a dimension that controls the difference between the two radii.

16. Place the dimension outside the larger radius and set the value to 5. Press Enter to accept the value, as seen in Figure 2.18.

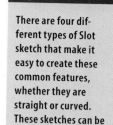

There are four different types of Slot sketch that make it easy to create these common features, whether they are straight or curved. These sketches can be used to create external shapes as well.

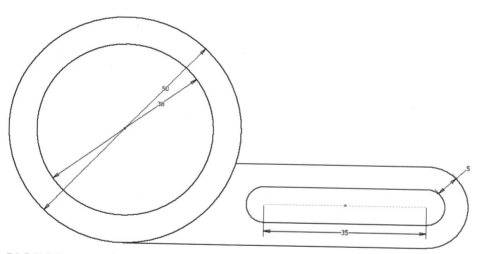

FIGURE 2.18 Dimensioning the difference between radii is a great way to control concentric circles.

17. End the Dimension tool.

It is easy enough to draw lines and arcs using constraints after the fact to control geometry, but the "in-line" options you used in this example allow you to keep focused on sketching without having to switch tools.

Making Sketch Entities More Than Powerful

Most of the sketching and constraints tools of Inventor are simple in their operation. Where there are techniques or shortcuts for getting more out of a tool the process for getting more is simple too.

Certification
Objective

1. Verify that the 2014 Essentials project file is active, and then open the c02-08.ipt file from the Parts\Chapter 02 folder.

   The visible sketch is made of two rectangles and two circles, and all are tangent so the lines and arcs are controlled by the two circles.

2. Press the E key to start the Extrude tool.

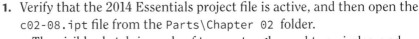

3. Move your cursor around to see that there are many profiles that can be used to create a solid modeling feature.

4. Press Esc to exit the Extrude tool.

   In this sketch the two circles are used to control the rectangles but should not be used for creating 3D features. There are many situations where any type of geometry can help to "construct" a sketch but are not needed. In these cases it is useful to declare these sketch elements as construction so they're not recognized by the 3D feature tools. As shown in this sketch, these construction elements can be constrained and dimensioned like regular sketch geometry.

5. Edit the sketch named Construction in the browser.

6. Select the two circles in the sketch and then pick the Construction override from the Format panel of the Sketch tab.

7. Change the small circle's diameter to **30** and the large circle's diameter to **150** to see how the diameters control the rectangles, as shown in Figure 2.19.

8. Finish the 2D sketch and start the Extrude tool.

   It is now easy to select the rectangles to create the proper 3D model.

9. Press Esc to cancel the Extrude tool.

10. Right-click the Construction sketch in the browser and deselect Visibility in the context menu.

11. Use the same steps to make the Centerline sketch visible.

    To revolve the sketch in its present state, you would need to select a profile and the axis for revolution. You will now make a modification to the sketch to simplify the modeling process as well as enhance the dimensioning on the sketch.

12. Edit the Centerline sketch.

You can start common 3D features like Extrude, Loft, Sweep, and Revolve using shortcut keys while still in the sketch environment.

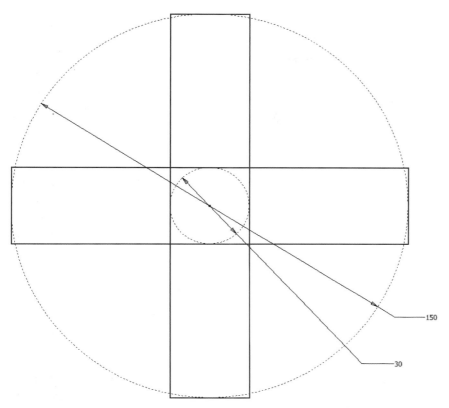

FIGURE 2.19 Controlling sketch geometry with construction geometry

13. Pick the line that sits alone in the sketch and then click the center-line override.

14. Start the Dimension tool from the Sketch tab's Constrain panel and pick the centerline and the lowest line on the sketch.
      Instead of a simple linear dimension, Inventor will preview a linear diameter dimension.

15. Place the dimension and set the value to 30.

16. Add the same type of dimension between the centerline and the top horizontal line; set its value to 80 and the middle line's value to 60 as shown in Figure 2.20.

◄

Lines that are part of the profile can also be made centerlines so that they are used as a revolve axis and linear diameter dimension reference.

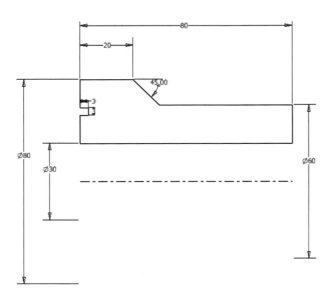

FIGURE 2.20 Sketch overrides can simplify dimensioning the sketch.

Fillet

17. Start the Fillet tool from the Draw panel of the Sketch tab.

18. Set the radius to **1 mm** and click the right vertical line and the horizontal line dimensioned to 60 mm.

19. Right-click and click OK on the marking menu to complete the fillet.

Chamfer

20. Use the pull-down menu option of the Fillet tool to start the Chamfer tool.

21. Set the dimension to **2 mm** and pick the bottom horizontal line and each of the two vertical lines to create two chamfered corners.

22. Click OK to finish the Chamfer tool.

23. Press the R key as a shortcut to starting the Revolve tool.
     This will select the profile and the axis and offer a preview of the revolved feature, as shown in Figure 2.21.

24. Press Esc to exit the Revolve tool and finish the active sketch.

There is no limit to the ways that construction geometry can be used. The centerline override adds a great option for those creating turned parts. You can change a linear dimension into a linear diameter dimension using a right-click option as well.

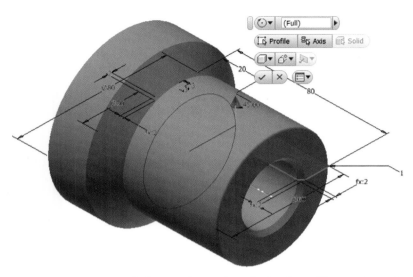

FIGURE 2.21 With the centerline in place, the Revolve tool knows what to do with the sketch.

Cleaning Up a Roughed-In Sketch

Extending and trimming geometry is a familiar concept for most 2D CAD users, and Inventor has these capabilities as well. Inventor also has other tools to help with making sure an imperfect sketch is ready for creating 3D features. Figure 2.22 shows how much simple editing tools can do.

Certification
Objective

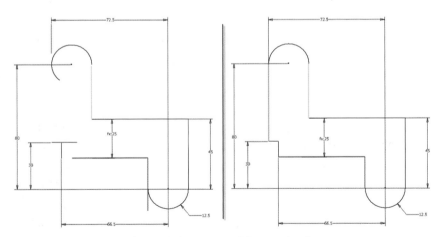

FIGURE 2.22 Before and after images of the part sketch

1. Verify that the 2014 Essentials project file is active, and then open the c02-09.ipt file from the Parts\Chapter 02 folder.

2. Edit Sketch1 by selecting the sketch geometry and choosing Edit Sketch from the mini-toolbar.

   There are a number of lines that are obviously disconnected. To assist in selecting the correct lines some of the ones that need editing have had their color changed.

3. Start the Extend tool from the Modify panel on the Sketch tab.

4. Pick the top of the yellow line and the left end of the green line to extend them to the next available boundary.

5. Hold the Ctrl key and click the short, red horizontal line as a boundary; then pick the vertical orange line to extend it. See Figure 2.23 for a reference.

> The Line Color, Line Weight, and Line Type settings of any sketch line can be changed by selecting the line, right-clicking, and selecting Properties from the context menu.

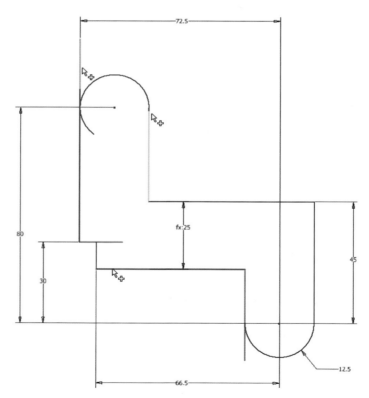

FIGURE 2.23 Extending in the sketch

6. Start the Trim tool from the Modify panel of the Sketch tab; then press the X key or, since Extend is still running, right-click and choose Trim from the context menu.

7. Pick the loose end of the arc, the top of the orange vertical line, the right end of the red horizontal line, and the bottom of the pink vertical line, as shown in Figure 2.24.

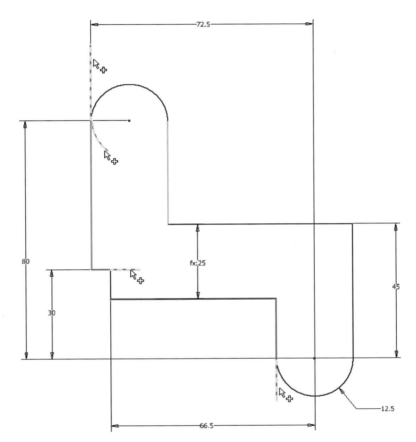

FIGURE 2.24 Trimming the sketch

8. Finish the 2D sketch from the Ribbon or from the marking menu. The sketch looks like it is cleaned up. If it is, it should be easy to extrude the shape.

9. Press the E key to start the Extrude tool.

The preview that is generated shows that the Extrude tool is trying to create a surface model. This is not what we need, so there is still a problem with the sketch.

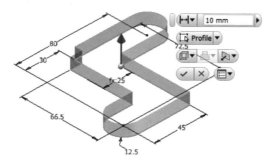

10. Expand the Extrude dialog box by selecting the bar at the bottom of the dialog box with the arrow on it.

11. In the Extrude dialog box, select the Solid option in the Output group.

    Inventor will highlight an icon in the dialog box with a red cross. This indicates that there is a problem with the sketch that is preventing it from creating a solid body.

12. Click the Examine Profile Problems icon in the dialog box.

13. The Sketch Doctor will open to the Examine page. On this screen (Figure. 2.25), a description of the problem will appear.

F I G U R E   2 . 2 5   The Sketch Doctor describes problems found in the sketch.

14. Click Next to select an option for solving the problem.

15. The Sketch Doctor will now offer ways to solve the problem. Select the Close Loop option and click Finish. When the Gap Between Points warning shown in Figure 2.26 appears, click Yes.

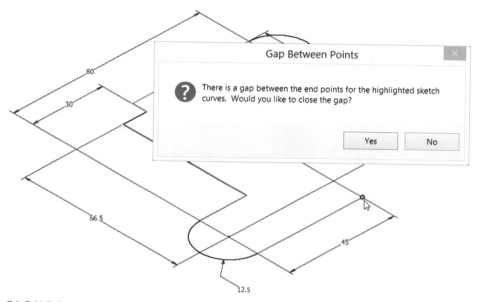

FIGURE 2.26 Inventor will highlight where a small gap exists in the sketch.

16. Click OK to close the confirmation that Close Loop was successful.

17. Press the E key again to start the Extrude tool and see that a solid feature is previewed in the Design window.

Having options to work with and repair sketch errors makes it much easier to get good-quality models.

Saving Steps in Sketching

Some complex sketches have duplicated geometry. Inventor has several tools to save you from having to create the same geometry over and over. Inventor also has tools for taking existing sketches and modifying their size as a unit.

Certification Objective

1. Verify that the 2014 Essentials project file is active, and then open the c02-10.ipt file from the Parts\Chapter 02 folder.

2. Edit Sketch1 by double-clicking Sketch1 in the browser.

3. Click the Scale tool in the Modify panel of the Sketch tab.

4. Drag a window around all of the sketched geometry to select it.

5. Pick the selection button for the Base Point and pick the lowest right corner.

6. Answer Yes to the prompts asking if you are willing to have Inventor relax existing dimensions and constraints.

7. Move your cursor to see how the sketch will be scaled, and then enter 1.5 in the Scale Factor field of the Scale dialog box, as shown in Figure 2.27.

FIGURE 2.27 It is possible to scale an entire sketch to a new size.

8. Click Apply to set the new scale and then click Done to close the dialog box.

9. Place a Fix constraint on any of the exterior sketch points.

   With the sketch scaled to a new size, it is time to replicate some of the features of the sketch. Inventor has Patterning tools for this that leverage parametric values to make editing them possible after they're created.

10. Select the Rectangular Pattern tool in the Pattern panel.

11. Click the pink circle, and then click the Direction 1 selector.

12. Click the green line; set the number of instances to 5 and the distance to 8 mm (see Figure 2.28).

13. Click OK to create the pattern.

14. Start the Circular Pattern tool.

FIGURE 2.28 Patterns can speed up sketch creation.

15. Click the other circle in the sketch and then click the Axis selector icon.

16. Pick the Center point of the diamond polygon to the right of it.

17. Set the number of instances to 5 and click OK to generate the pattern shown in Figure 2.29.

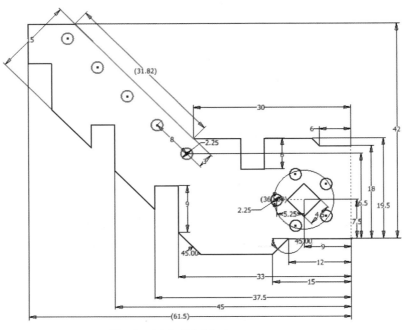

FIGURE 2.29 The sketch is half finished.

The last task is to mirror the sketch.

18. Start the Mirror tool from the Pattern panel on the Sketch tab.

19. Drag a window around all of the geometry.

20. After picking the geometry, click the Mirror line selection icon and then pick the construction line on the right side of the sketch.

21. Pick Apply, then Done to complete the sketch and finish the tool. The mirrored sketch not only saves time but will behave with symmetry.

22. Double-click the vertical dimension with the value of 42 and change it to **45**.

23. Find the distance dimension of the rectangular pattern and change it from 8 mm to **9 mm**.

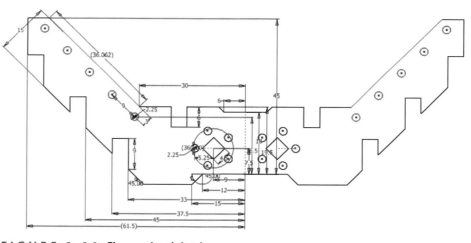

Dragging a window from left to right requires completely encompassing geometry for selection. Dragging from right to left will allow you to cross geometry only in part to select it.

Your finished sketch will look like Figure 2.30. Because a construction line defines the mirror line, it can be selected with a single click to create an extrusion. You can now close the file without saving.

FIGURE 2.30 The completed sketch

Creating Splines

Splines offer you a huge amount of control for creating complex shapes. Inventor offers two types of splines offering different ways of controlling their shape. The two common approaches are Interpolation and Control Vertex.

These steps will just scratch the surface of this powerful geometry type but will give you a good foundation.

1. Verify that the 2014 Essentials project file is active, and then open the c02-11.ipt file from the Parts\Chapter 02 folder.

2. Start the Create 2D Sketch tool from the Sketch tab and select the XY plane under the Origin folder.

    The Interpolation technique is the technique most commonly used in engineering. You simply pick points directly along the path you want the spline to follow and it calculates how the spline should pass through the points. These points can be controlled in a number of ways after they've been placed to refine the final shape of the spline.

3. Start the Spline Interpolation tool from the Draw panel.

4. Pick a series or points following the top curve.

5. Click the green check mark on the mini-toolbar to create the spline as shown in Figure 2.31.

Spline
Interpolation

A curvature control can be applied to a point to define a radius that the spline will transition across along with the direction and weight.

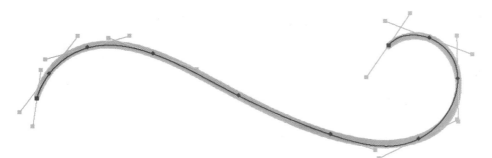

FIGURE 2.31   Splines created with Interpolation use points along a path.

After the spline is created, a series of lines will appear through the points. These are handles, and their length applies a "weight" that controls how close to the point the curvature transitions. The value of this weight can be controlled with a parametric dimension. The angle that the curve transitions through can also be controlled through a dimension, and the point itself can be dimensioned into position.

6. Start the Spline Control Vertex tool from the Draw panel.

Spline
Control Vertex

The Control Vertex technique still uses points, but they are placed off the desired path with the spline passing between them. This technique is more common among industrial designers and might be considered more artistic.

You can select a spline and insert a point or vertex to add additional control if needed.

7. Experiment with placing points using the bottom curve for a guide until you achieve a satisfactory result. See Figure 2.32 for suggestions.

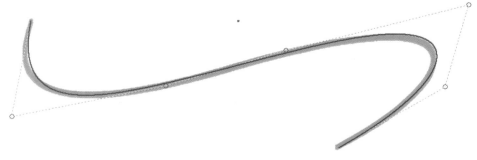

FIGURE 2.32   Control vertices sit off the spline.

8. Right-click and select Create from the marking menu to complete the spline.

The points placed for the Control Vertex Spline can also be controlled by dimensions, but only their position can be altered.

An entire chapter could be spent exploring the Spline tools, but they are not as commonly used as other sketch elements. If you need to create complex curvature, it will not take you long to master the technique that works best for you.

Using Sketches for Concept Layout

Certification Objective

Using 2D to test a concept is not new. Using a combination of parametric dimensions and sketch constraints can bring this idea up-to-date and offer quick changes to test an idea.

1. Make certain that the 2014 Essentials project file is active, and then open the c02-12.ipt file from the Parts\Chapter 02 folder.

2. Double-click the Sketch icon in the browser to activate the Linkage sketch.

Under the sketch in the browser, you will see two instances of a sketch block named Link 1 and one instance of a sketch block named Base. Two circles and a line also reside in the sketch but are not part of a sketch block.

3. In the Layout panel of the Sketch tab, click the Create Block tool. Adding entities to a sketch block allows you to group the sketch elements together without creating a series of sketches.

    Create Block

4. Click the pink circles and line.

5. Pick the Insert Point select tool, and then pick the center of the right circle.

6. Enter **Link 2** for the block name, and click OK to create a new sketch block.

7. Click the center point of the circle on the right of the new block, and drag it to the center point at the top of the vertical red bar on the right (see Figure 2.33); release when it shows it is creating a constraint between centers.

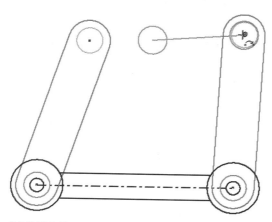

FIGURE 2.33 Dragging sketch constraints between sketch blocks

`8. Click and drag the center of the new block on the left to the center point of the vertical bar on the left.

9. Click and drag one of those endpoints to see how the linkage moves.

10. Double-click one of the instances of Link 1 in the Design window.

11. When the sketch geometry appears, double-click the 60 mm dimension and change it to **80 mm**.

Finish Edit Block

12. Click the Finish Edit Block icon on the Exit panel of the Sketch tab.

13. Move the links using the new part length.

Sketch blocks can be simple or complex as needed to help you predict how a mechanism will work. Sketch blocks are quick, and the sketch can be used to build 3D data and speed up the development of an assembly.

THE ESSENTIALS AND BEYOND

The "correct" way to create a sketch is a very personal thing and is based on the needs for updating the model and whether it is likely that its overall shape may need to change and how.

ADDITIONAL EXERCISES

▶ Work with creating shapes that you need using primitives rather than sketching in the perimeter of the shape.

▶ Explore the arc options available in Inventor.

▶ Use Construction geometry to create the shapes of parts that your active part will connect with in the sketch to define the part.

▶ Work on mixing and matching dimension units in dimension values, including making statements like **1in - 5mm** to see that Inventor can calculate the value.

CHAPTER 3

Introducing Part Modeling

Part modeling is typically the primary focus for Autodesk® Inventor® beginners. The tools involved are the ones you will spend a majority of your time using, but they are pretty straightforward.

In this chapter, you will look at the most commonly used tools for creating 3D solid models.

In short, the process of beginning a new part consists of creating a sketch that defines the most important elements of the part's geometry and then turning that sketch into a 3D shape. Along the way, you add features that complete the part. That really is how simple the process is.

▶ **Creating 3D geometry: The parametric solid model**

▶ **Defining the base feature**

▶ **Creating sketched features**

▶ **Defining axial features**

▶ **Building complex shapes**

▶ **Including placed features**

Creating 3D Geometry: The Parametric Solid Model

The most basic type of 3D feature in Autodesk Inventor is the sketch feature. Sketch features depend on the existence of a 2D sketch or sketches to define them. This sketch foundation also allows them to create extraordinarily diverse geometry. In Chapter 2, "Introducing Parametric Sketching," you created sketches that could be used to create 3D features. The power of these parametric sketches lies in the fact they can be edited after they've been used and update the 3D geometry based on them.

Combining the various features will build the parts you need, including the one you will build in this chapter, shown in Figure 3.1.

FIGURE 3.1 The focus of this chapter will be to create this part.

 N O T E   The following modeling exercises describe the procedures using the mini-toolbar and in-window tools. All the tools and options are also available in the Ribbon or in dialog boxes.

Defining the Base Feature

Certification
Objective

One of the great challenges of defining and creating a 3D shape is deciding where to begin. The first feature you'll create is referred to as the base feature. Think about the process of creating a solid model as constructing it.

Starting with a solid foundation is important for all construction. The base feature of your part should either define much of the part's primary shape or create a feature that all other features are built on.

Using the Extrude Tool

An extrusion takes a 2D profile and extends it a distance perpendicular to the plane of the sketch. It can be set to a specific distance, evenly divided in two directions at once, given unique distances on either side of the sketch plane, or terminated at or between faces or planes in the part.

In this exercise you will use the Extrude tool to define the base feature of the part:

1. Make certain that the 2014 Essentials project file is active, and then open c03-01.ipt from the Parts\Chapter 03 folder at this book's web page: www.sybex.com/go/inventor2014essentials.

2. In the Design window, click one of the sketch lines. As you do so, commonly used tools will appear.

3. Click the Create Extrude tool on the left.

4. An initial preview of the extrusion will appear with a mini-toolbar in the middle of the screen and a full dialog box. Both show the current value for the extrusion distance.

5. Click and drag the arrow at the end of the extrusion until the value in the mini-toolbar reads 15 (see Figure 3.2). Or you can highlight the value in the mini-toolbar and enter **15**.

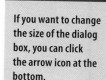

If you want to change the size of the dialog box, you can click the arrow icon at the bottom.

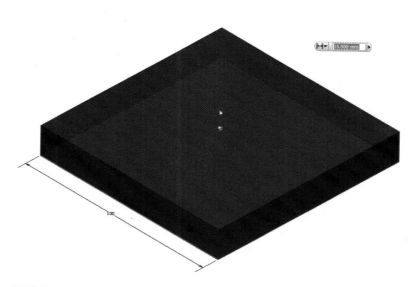

FIGURE 3.2 The option in the dialog box updates while you drag the direction arrow.

6. On the mini-toolbar, click the green check mark (which represents OK) to create the feature.

The extrusion is generated on your screen, and a feature named Extrusion1 appears in the Browser.

7. Close the file without saving.

This simple action is the foundation of 3D parametric solid modeling. These steps and similar steps for other sketched features will be repeated over and over to create any type of 3D solid model with Inventor.

Options for Starting a Part

A part with a complex primary shape will simply need a more complex sketch. When you have a simple part, you have several options for creating a simple shape.

For example, the extrusion you created in the previous exercise was based on a sketch that was constructed using a series of lines or a Two Point Rectangle. For this exercise, you will use two different options to create a simple sketch, but there are many ways to create the correct 3D geometry.

1. On the Quick Access toolbar, click the New icon.

2. Make sure that 2014 Essentials.ipj is listed as the active project file near the bottom of the dialog box.

3. In the New File dialog box, select the Metric Template set on the left.

4. Locate and double-click the Standard (mm).ipt template file.

5. With the new file open, right-click and select New Sketch from the marking menu.

6. In the Browser expand the Origin folder and pick the XZ plane to start a new sketch on.

7. Start the Polygon tool from the Draw panel of the Sketch tab.

8. Change the number of sides to 4 and pick the projected center of the part as the origin.

9. Move the cursor on the screen to see that it is drawing a square centered on the origin. Pick somewhere in the Design window to create the new geometry

10. Start the Horizontal constraint from the Constrain panel of the Sketch tab; then select one of the lines to square the sketch with the axes of the part but do not add a dimension.

11. Press the E key to start the Extrude tool and create a solid 15 mm tall.

    It is not necessary to have a completely constrained sketch to create 3D features in Inventor as long as the sketch is a completely closed loop.

    The geometry created using the polygon is the same as the first feature you created in this chapter. That shape is essentially a box, which is a primitive shape. Primitive creation tools automate the process of creating the sketch and the sketched feature.

12. In the Browser right-click on the newly created Extrusion1 feature and click Suppress Features in the context menu to remove the Extrusion feature from the Design window.

    The feature has not been deleted. Its properties are still stored in the file; it is simply not generated.

13. Find the Box tool on the 3D Model tab and start it from the Primitives panel.

    The tool requires that you set a plane to build your feature from.

14. Pick the XZ Plane under the Origin folder.

15. A new sketch will be created. Click the projected center point of the part as the center point for the sketch.

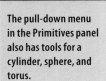

The pull-down menu in the Primitives panel also has tools for a cylinder, sphere, and torus.

16. After picking the center, move the cursor away from the center and you'll see that the Center Point Rectangle tool has been started. You can drag the size of the rectangle using the automatic dimensions as a guide.

17. Enter **100** for the value of the first dimension, press the Tab key, and enter **100** for the value of the second dimension.

18. Press Enter to complete the sketch and begin to drag the height of the Box feature.

19. After the view repositions, set the height of the box to **15**, right-click, and choose OK from the marking menu to create the feature.

20. Pick a face of the feature and select Edit Sketch from the toolbar to see the sketch with dimensions and construction lines shown in Figure 3.3.

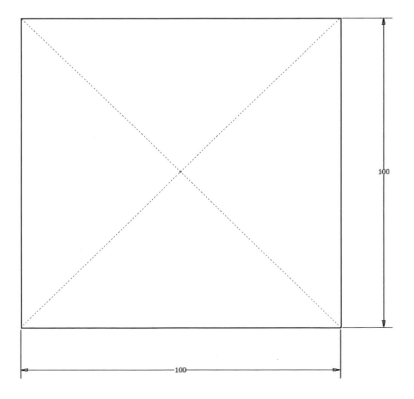

FIGURE 3.3 The automatically generated sketch at the heart of the Box primitive

21. Close the model without saving.

In the Browser an Extrusion2 feature was created. The Box tool automates the process of sketching and extruding a primitive shape.

Setting the Material and Color

Creating a computer representation of your design makes it easy to understand form and function. You can also do things such as establish the material your part is made from and monitor its weight or inertial properties. You can even pick a specific color if you like.

1. Make certain that the 2014 Essentials project file is active, and then open c03-02.ipt from the Parts\Chapter 03 folder.

2. In the Browser, right-click the part icon at the top, and select iProperties from the context menu.

3. In the dialog box, go to the Physical tab, and use the Material drop-down to select Iron, Cast as the material. Notice that doing so updates the mass; make note of that value.

4. Click OK to accept this change.

5. In the Quick Access toolbar, locate the drop-down that shows Generic for the part's color. Select a color you like, and feel free to change it whenever you want.

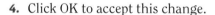

At various times while working in this and other chapters, you should go back to the Physical tab and update the properties to see how changes affect these values.

Setting the material from which the parts will be made from is important for estimating costs, weight, and so on. Color changes can help highlight components or make them appear more realistic.

> **TIP** Inventor also features a few different color libraries. If you intend to work with other Autodesk applications like Autodesk® 3ds Max® and you want to maintain consistency in the color, you should use a color from the Autodesk Appearance library.

Reusing Sketch Geometry

On a complex part where there are many features created from the same base plane, you can sometimes use a single sketch to define more than one feature.

Certification Objective

1. Continue using the drawing from the previous exercise, or make certain that the 2014 Essentials project file is active and then open c03-03.ipt from the Parts\Chapter 03 folder.

2. Click the small + icon in the Browser next to Extrusion1. This exposes the sketch that was consumed by that feature.

3. Right-click the sketch, and select Share Sketch from the context menu.

   The sketch now appears both under and above the extrusion feature. It also becomes visible in the Design window so other features can use it. In this case, you need to add to it first.

4. Click the top face of the extrusion. When you do, icons will appear for Edit Extrude, Edit Sketch, and Create Sketch.

5. Click the Edit Sketch icon.

6. The Center Point Circle tool is on the marking menu and in the Draw panel. Start it, and draw a circle based on the part center point with a diameter of 40.

7. Start the Offset tool on the Modify panel of the Sketch tab, select the circle, and pick on the outside to make a larger circle. Press Esc to stop the Circle tool. See Figure 3.4.

Your sketch might rotate so that you're looking directly at it. To create images in an isometric view, I have disabled this feature in the application options to show that sketches can be edited in this way as well.

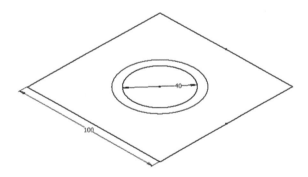

FIGURE 3.4 Adding new geometry to an existing sketch

8. Start the Dimension tool, and click the two circles. This will create a dimension showing the distance between the perimeters. Set it to 10, and finish the Dimension tool.

9. Right-click and select Finish 2D Sketch from the marking menu.

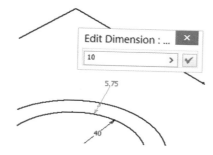

Now you can create more sketch geometry that is based on the center of the part. This is why it can be useful to think about centering a part on the origin planes.

1. Continue using the drawing from the previous exercise, or make certain that the 2014 Essentials project file is active and then open c03-04.ipt from the Parts\Chapter 03 folder.

2. Expand the Origin folder in the Browser.

3. Click the XY plane. This will highlight the plane in the Design window and present you with the Create Sketch icon.

4. Click the icon to start a new sketch on that plane.

5. The center point of the part will automatically project into the new sketch, but it will be difficult to see it from the front.

6. Right-click in the Design window, and select Slice Graphics (or press the F7 key) to remove some of the part to expose the new sketch. You can also pick the Slice Graphics icon in the status bar.

7. Start a new line at the part's center point; going straight up, click roughly at 30 mm.

8. Move your mouse to the endpoint of the line you just created, click, and drag an arc tangent to the last line, as shown in Figure 3.5.

9. Add a straight line that is tangent to the arc. A glyph will appear that shows when the line is tangent. Press Esc to finish the Line tool.

10. Select General Dimension from the marking menu.

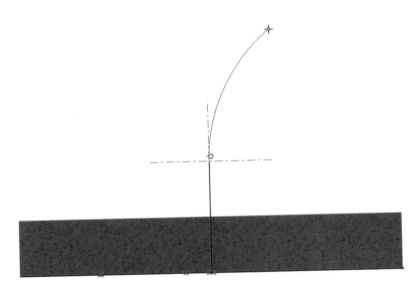

FIGURE 3.5 A tangent arc can be created using the Line tool.

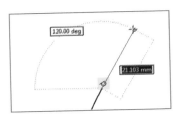

11. Click the arc. When the radius dimension preview appears, right-click, and from the Dimension Type menu select Arc Length. Place the dimension and set the value to **50**.

12. Click the second straight line, right-click, and click Aligned. Place the dimension and set the length of the line to **30**.

13. Now click the first line and then the second line. Because they are not parallel, an angular dimension will appear. Position it as shown in Figure 3.6, and set its value to **30**.

14. Finish the sketch.

Getting feedback on constraint conditions like tangency during the sketching process helps you know you're getting the control you need from the beginning.

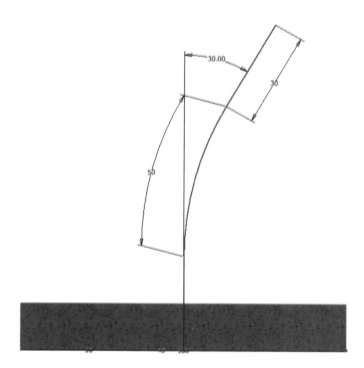

FIGURE 3.6 Refining the sketch with parametric dimensions

Connecting to Other Sketch Data

Linking to geometry in other sketches helps keep things in line.

1. Continue using the drawing from the previous exercise, or make certain that the 2014 Essentials project file is active and then open c03-05.ipt from the Parts\Chapter 03 folder.

2. Start a new sketch on the XY plane.

3. Use the Project Geometry tool on the marking menu or in the Draw panel, and click the top 30 mm line to link it into the current sketch.
   Read the next several steps carefully and refer to the graphics to help understand the steps that you will carry out.

4. In the Draw panel, expand the Rectangle tool, and start the Three Point Rectangle tool.

5. Start the rectangle at the top end of the projected line. Drag to the right so it is perpendicular to the projected line, and click.

6. Place the third point above the rest of the sketch at approximately 10 mm, as shown in Figure 3.7.

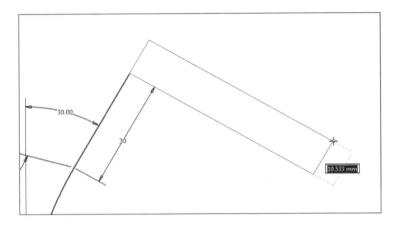

FIGURE 3.7 A three-point rectangle can be placed at any initial angle.

7. Stop the Rectangle tool by pressing Esc or by selecting Cancel from the marking menu.

   The new rectangle will eventually be revolved so that even though you are looking only at the side, it would be best to dimension it as its final diameter. Using the Centerline override, you can get the dimension tool to recognize this.

8. Click the short edge of the rectangle that is collinear with the 30 mm line, and then pick the Centerline override in the Format panel.

9. Start the Dimension tool, and first click the centerline and then the opposite end of the rectangle.

10. Place the dimension, and set its value to **110** mm. See Figure 3.8.

11. End the Dimension tool; then drag the top edge of the rectangle down so it is positioned below the end of the 30 mm line.

12. Restart the Dimension tool, and place a 10 mm dimension for the thickness by picking long lines at the top and bottom of the rectangle.

13. Finish the Dimension tool and editing the sketch.

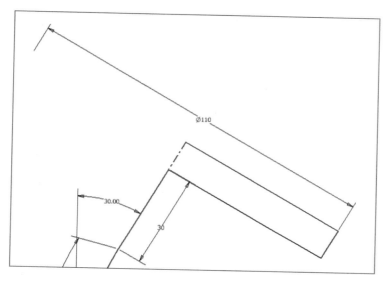

FIGURE 3.8 A linear diameter can be placed in the sketch.

Let's keep building on your part. Moving forward, you will be switching files because some of the basics will be prepared for you so you can focus on new things.

Defining Axial Features

A Revolve feature requires a 2D profile and an axis for this profile to revolve around to define a 3D feature. The termination options are similar to an Extrude feature using distances or faces to define the extents of the revolution.

Creating a Revolve Feature

For many users, revolving is the second most common thing to create with a sketched feature after extruding. If you make turned components, it may be your number-one tool, although I think you'll like the Shaft Generator covered in Chapter 9, "Advanced Assembly and Engineering Tools."

Certification
Objective

In this exercise you will use a couple of interesting sketching techniques to set up a Revolve feature.

1. Make certain that the 2014 Essentials project file is active, and then open c03-06.ipt from the Parts\Chapter 03 folder.

2. Click an element of the rectangle in Sketch3, and choose the Create Revolve icon.

3. Because there are multiple visible sketches with more than one complete profile, you must click the shape of the rectangle.

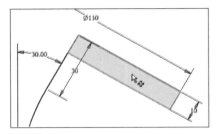

4. Click the Axis icon on the mini-toolbar to let Inventor know you're done selecting profiles.

5. Select the centerline that is in line with the 30 mm dimension.

   This will immediately produce a preview of the new feature (Figure 3.9). Note that an arrow allowing for the changing of the angle is available.

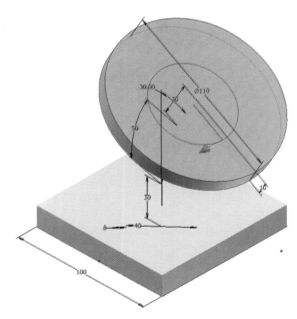

FIGURE 3.9 A preview of the Revolve tool

6. Finish the Revolve tool by clicking the green check mark.

Now you can complete the main features of the part.

Building Complex Shapes

Extrusions take shapes and add a thickness to them in a single direction. Sweep features can define a more complex shape.

To create a sweep you need a 2D profile and at least one additional sketch for that profile to define a rail to follow. It acts very much as an "extrusion along a path" and can include a third sketch or use an existing surface to further control how the shape follows that path.

Creating Sweep Features

In this exercise, you'll create two features. The first will establish the outer portion and will be hollow. The second will cut through the top and bottom plates.

1. Make certain that the 2014 Essentials project file is active, and then open c03-07.ipt from the Parts\Chapter 03 folder.

2. Start the Sweep tool from the Create panel on the 3D Model tab.
Adding the circles to Sketch1 has made it a multiloop sketch, so there is more than one option; the button for profile selection in the dialog box will have a red arrow. This means Inventor needs you to make a selection.

3. Click in the space between the concentric circles.

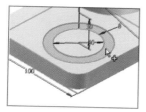

4. Because Optimize For Single Selection is selected, Inventor now wants you to click the path. Select the part of the lines and arc going to the disk on top, as shown in Figure 3.10.

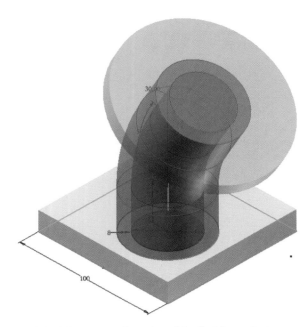

FIGURE 3.10 A preview of the first Sweep tool

5. Click OK in the dialog box to finish the tool.

   This consumed the sketch that you used for the sweep path. When a sketch is consumed it disappears from the screen once it has been used to create a 3D feature.

6. Expand the Sweep feature in the Browser.

7. Right-click and make Sketch2 visible.

8. Start the Sweep tool, and use the inner circle for your profile.

9. Select the same path sketch.

10. Set your option in the dialog box to Cut.

11. Click OK to create the feature.

12. Turn off the visibility of the two sketches. Figure 3.11 shows the part at this point.

By sharing a sketch that defines more than one feature, you can control the relationships between the various sketches. Now you will focus on adding a few placed features to complete the part.

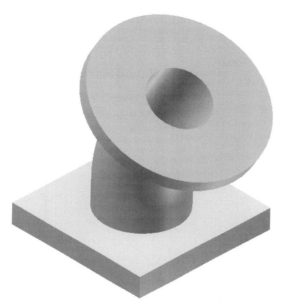

FIGURE 3.11 Most of the part is finished.

Leveraging Primitives

So far, you've created simple shapes such as boxes and cylinders using sketched features, but that was more about sketching than creating solid models. Primitive shapes such as boxes, cylinders, and spheres are easier to place as intact features. You might even want to re-create the model using these primitives to explore that option. For now, you will use one to add geometry to the model.

1. Make certain that the 2014 Essentials project file is active, and then open c03-08.ipt from the Parts\Chapter 03 folder.

2. Use the menu in the Primitives panel to start the Cylinder tool.

3. Select the top face of the revolved feature for a plane to place the cylinder on.
   This will automatically project the geometry of that face, including the center point of the concentric circles.

Cylinder

4. Pick the center of the circles for the center of the cylinder and drag the diameter.

5. Enter a diameter of 55, and press Enter.

Doing so automatically switches the view position of the model and starts the Extrude tool.

6. Drag the distance arrow down and back into the part to convert the operation from a join to a cut, as shown in Figure 3.12.

7. Set a depth of 5 mm, and click OK to place the new feature.

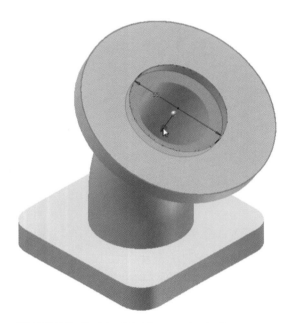

FIGURE 3.12 Primitives accelerate the process of creating basic shapes.

Looking at the Browser, you won't see a special feature callout, and editing the extrusion that was created is no different than if you had gone through the process step by step. As with many of the most productive tools in Inventor, this is just an easier way to do things that allows you to stay focused on what you're creating, not on how you're creating it.

Including Placed Features

Placed features modify geometry that is already on the part and do not require sketches to define them. At least one of these features will exist on nearly every part.

Adding an Edge Fillet

One of the best practices for modeling is *not* to include incidental fillets in the sketch. It is better and easier to place fillet features in the model instead.

Certification
Objective

1. Continue using the drawing from the previous exercise, or make certain that the 2014 Essentials project file is active and then open c03-09.ipt from the Parts\Chapter 03 folder.

2. Click one of the short, vertical edges on the square base of the part.

3. When the icons appear for the most common options, select the Create Fillet tool.

4. Click the remaining three short edges.

5. An arrow icon will follow your selections. Drag it to create a 14 mm radius on the four corners, as shown in Figure 3.13.

6. Thanks to marking menus, a quick right-click and drag to the right will give you the same effect as clicking OK. Try this to complete the Fillet tool.

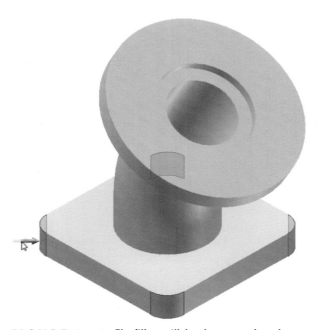

FIGURE 3.13 The fillets will develop as you drag them.

Fillets are very common, and in Chapter 7, "Advanced Part Modeling," I cover additional techniques for placing them. It is also possible to place fillets with several different values at the same time.

Adding Edge Fillets with Multiple Radii

Fillets are another critical feature in machine parts. Fillets have many options, but for this chapter, you will focus on the edge fillet and placing fillets of more than one radius using a single feature.

It is possible to click faces and edges through a part by hovering over their location. You can also orbit while in a command without interrupting it.

1. Make certain that the 2014 Essentials project file is active, and then open c03-10.ipt from the Parts\Chapter 03 folder.

2. Click the round edge where the sweep meets the extrude.
   This presents a choice of the Create Fillet or Create Chamfer tool.

3. Select the Create Fillet tool.

4. Add the edge where the sweep feature meets the revolved feature.

5. Drag or enter a radius value of **6**.

6. In the Fillet dialog box, click where the words *Click to add* appear. Doing so adds a new row for using another radius size.

7. Click the radius value of the new row, and change it from 6 to 2.

8. Click the outer edges of the extrusion and revolved features near the first set of fillets. See Figure 3.14.

9. Click OK to finish applying the fillets.

The option of placing a fillet or a chamfer on a sharp edge makes perfect sense, so it is great that you have access to the option by simply picking a sharp edge.

FIGURE 3.14 You can place fillets with different radii in one fillet feature.

Applying a Chamfer

Chamfers and fillets have similar locations, but their options are quite different. For this exercise, you will use two of the three options:

Certification Objective

1. Make certain that the 2014 Essentials project file is active, and then open c03-11.ipt from the Parts\Chapter 03 folder.

2. Click the sharp edge on the interior of the revolved feature.

3. Select the Create Chamfer tool. The default is 45°. Drag the Chamfer value to 2 mm, as shown in Figure 3.15.

4. Finish the chamfer.

5. Rotate the part so you can see the bottom.

6. Select the sharp inside edge, and select the Chamfer tool.

7. Set the option to Two Distances (you may have to reselect the edge).

FIGURE 3.15   Applying a chamfer to the model edge

8. Set the first value to 2 and the second distance to 4.

9. Finish the chamfer. See Figure 3.16 for the finished adapter.

Now you can add the final touches to your part by placing holes.

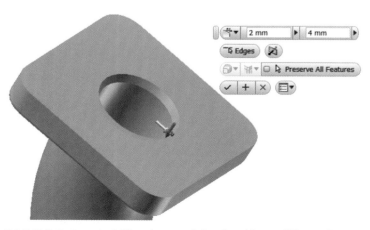

FIGURE 3.16   Adding the second chamfer with two different sizes

The Hole Feature

One of the most common features of machined components is holes. Inventor has a variety of hole classes (drilled, counterbore, spotface, and countersink) as well as a number of ways to position and size them.

Linear Hole Placement

Hole features are critical geometry, and they are defined and positioned in a variety of ways. The Hole dialog box offers a single location for all of the options for the Hole tool.

Certification
Objective

1. Make certain that the 2014 Essentials project file is active, and then open c03-12.ipt from the Parts\Chapter 03 folder.

2. Start the Hole tool from the marking menu or the Modify panel of the 3D Model tab.
   The default options for hole placement (when no visible sketch is available) is to create a drilled hole to a size you specify located on a face and positioned off of up to two edges.

Hole

3. Pick the top face of the box feature at the base of the part.

4. In the dialog box, set the Termination option to Through All and the Diameter to **10 mm**, as shown in Figure 3.17.

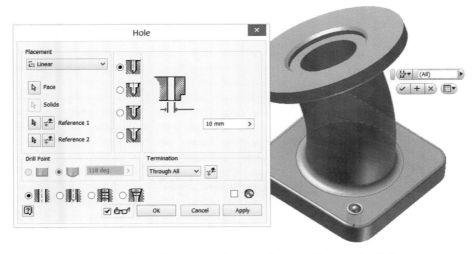

FIGURE 3.17 A Linear placement can be approximated without setting Reference dimensions.

5. Click and drag the center point of the hole to see that it can be moved around the face.

    To properly locate a linear hole, you need to select two reference edges. These edges do not have to be perpendicular to one another, and they do not have to be on the face where the hole is placed.

6. Pick the selection icon for Reference 1 in the Hole dialog box.

7. Use the ViewCube® for direction reference and pick the back edge of the bottom face and set the value to **30 mm**.

8. Pick the Reference 2 icon and pick the bottom-left edge of the part with an offset 10 mm.

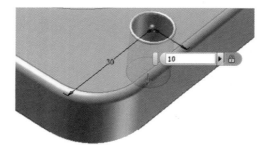

9. Click OK in the Hole dialog box to create the hole.

The hole can also be placed to a specific depth or placed to reach a selected face. These, along with other size options and hole styles, will be explored in the next exercises.

Placing Concentric Holes

Certification
Objective

Another common hole placement is centered on a rounded feature. A Concentric hole is located by a radiused edge or face.

1. Make certain that the 2014 Essentials project file is active, and then open c03-13.ipt from the Parts\Chapter 03 folder.

2. Start the Hole tool from the marking menu or the Modify panel of the 3D Model tab.

3. In the Hole dialog box, set the placement to Concentric.

4. Select the top face of the bottom extrusion as the plane for the hole to begin.

5. Click the nearest 14 mm fillet face for the concentric reference.

6. Drag the hole's diameter by clicking and dragging the ring that appears on the face to see how this can be approximated.

7. Between the hole placement tools and the preview of the hole geometry, specify the hole class as Countersink.

8. Set the hole type to Clearance for an ISO Standard Countersunk Raised Head Screw ISO 2010/7047 with an M10 thread and Normal fit. See Figure 3.18.

9. Make sure the Termination setting is Through All.

10. Click OK to place the hole feature.

You can make the bolt pattern across the bottom plate in several ways. For instance, you can repeat the Hole command for each corner or do a rectangular pattern. Or, because the first feature is centered on the z-axis of the part, you can use a circular pattern, as you'll see in the next section.

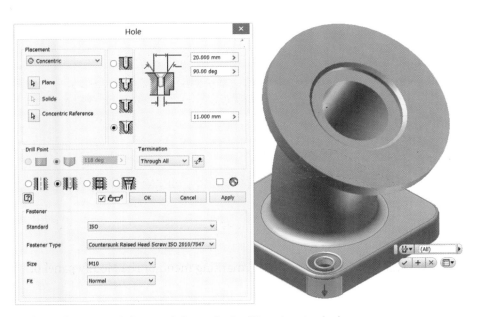

FIGURE 3.18 A clearance hole can size itself based on standards.

Creating a Circular Pattern

Now you will make three more holes the easy way, and they will be associative to any change to the original hole:

1. Make certain that the 2014 Essentials project file is active, and then open the c03-14.ipt file from the Parts\Chapter 03 folder.

2. Select the Circular Pattern tool from the Pattern panel of the 3D Model tab.

3. In this case, you only need to select the Hole2 feature for the features selection.

4. In the dialog box, click the Rotation Axis button.

5. Expand the Origin folder in the Browser if needed, and select Y Axis; you can also pick the lowest cylindrical face of the sweep feature.

6. Set the value in the Placement field to 4, and check the preview.

7. When the preview shows new holes added to the other corners, click OK to create them. Figure 3.19 shows the part at this point.

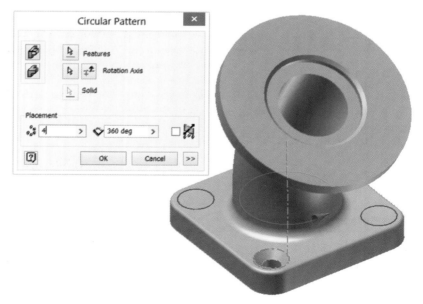

FIGURE 3.19 Using a pattern saves on repetition and helps prevent mistakes.

You might also choose to use a rectangular pattern to create this feature. The circular pattern would remain correct if the footprint of the box feature changed size.

Placing Sketched Holes

To create complex patterns of holes, you can develop a sketch that includes a number of hole centers or naturally occurring points. By selecting them, you can place many holes in a single feature.

Certification
Objective

1. Make certain that the 2014 Essentials project file is active, and then open c03-15.ipt from the Parts\Chapter 03 folder.

   The dotted lines and circle that appear in the sketch have an override applied to them: Construction. This prevents them from being recognized as profiles by the 3D features but allows them to be used to construct other things. In this case, you'll use them for the location of the center points for holes. The corners of the square have an override applied as well. This will make the Hole tool find them automatically.

2. Start the Hole tool and set the type to Drilled. As mentioned, four locations will be selected.

3. Set the type to Tapped Hole, the thread type to ISO Metric profile, and the size to 10. Select the Full Depth check box for the threads.

   When creating holes, you can set a depth for the threads that's different from the pilot hole. Using the Full Depth option bypasses the need to specify a thread depth.

4. Change the termination to To, and click the topmost cylindrical face so the holes stop there. See Figure 3.20.

5. Click OK to complete the hole.

You've now used nearly every option of the Hole feature. Since you are this close, let's see what these last options are.

Placing a Hole on a Point

The most complex hole placement option is On Point. To use this placement you need a work point feature and a work axis. This is a more specialized option well suited for creating hole features that are not perpendicular to a plane.

Certification
Objective

FIGURE 3.20 Hovering over an obscured object will eventually cause a drop-down to appear, allowing you to select it.

1. Make certain that the 2014 Essentials project file is active, and then open c03-16.ipt from the Parts\Chapter 03 folder.

2. Start the Hole tool from the marking menu and change the class to Spotface.

3. Change the Placement option to On Point

4. Pick the work point where the axis meets the face of the part.

5. Inventor will automatically select the Direction option, so click the axis in the part.

6. Set the bore diameter of the hole to **20 mm** and the bore depth to **2 mm**.

7. Set the hole type to Taper Tapped Hole, the thread type to NPT, and the size to 1/8.

8. Set Termination to To and select the inner face of the sweep that makes up the neck of the part. You may need to hover over the outer face and highlight the inner face using the pull-down menu to select it, as shown in Figure 3.21.

9. Click OK to create the feature and complete the part.

The Counterbore and Spotface tool create the same geometry but are dimensioned differently. The bore of a spotface is used to create a face on an uneven surface to measure the depth of the hole from.

In later chapters you will learn to create the work features used to complete this part.

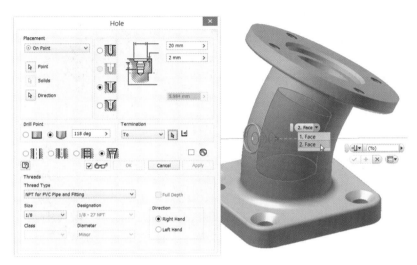

FIGURE 3.21   Even selecting several options is easier than using extrudes and revolves to create holes.

THE ESSENTIALS AND BEYOND

There are many more options in part modeling that will be covered in later chapters. In this chapter, you worked with the basic tools that will allow many people to do most of the things they need to do. You may also have noticed that the tools do a good job of prompting you for the things they need to help you.

ADDITIONAL EXERCISES

► Try modeling the base with the Polygon tool.

► Explore whether using an extruded circle is more appealing to you than using the Revolve tool.

► Try modifying the angular dimension on the sweep path sketch to see the effect.

► Experiment with various types of holes to see how they change with different fasteners.

CHAPTER 4

Creating 2D Drawings from 3D Data

For those new to Autodesk® Inventor® 2014, it is easy to focus entirely on the 3D design tools. However, many users still need to produce 2D production drawings.

Chances are, if you're reading this book, you have created detail drawings and will likely need to create them in the future. Creating detail drawings from the solid (or surface) model is so easy it can often be considered fun.

To create a new drawing in Inventor, you don't have to have a 3D file open. In fact, if your system resources are limited, it is best not to have the file open.

▶ **Drawing views of a part**

▶ **Editing views**

▶ **Adding detail to drawing views**

▶ **Dimensioning**

Drawing Views of a Part

When you've created a drawing view, it must accurately reflect the model geometry. If you modify the geometry, that change should be reflected in all drawing views. This is just standard drafting practice, and the way Inventor works as well, but you may notice things happen a bit more quickly than on the drawing board.

This section introduces the types of drawing views and then provides exercises where you will create and work with views.

Types of Drawing Views

Any type of drawing view that you've created on the drafting board or in a 2D CAD system can be created with Inventor. The process is a lot faster in Inventor because you're generating views of the object(s) rather than generating a lot of separate pieces of geometry to represent the same edge or feature. Here are the types of drawing views in Inventor:

Base View   If you think about how you currently do drawing layouts, you are already working with a hierarchy, whether or not you're doing so consciously. You have a view that others are positioned around. In Inventor, this is referred to as the *base view*, and there can be as many base views as you need or as many components as you like, all on one page.

Projected Views   Once a base view is placed, you can immediately place standard orthographic projections and isometric views from that base view. Any change that is made to the scale, position, or contents of the base view will be reflected in the projected view by default.

Section Views   These are placed much like projected views but with a section line defining what geometry will be generated. The section line can have multiple segments or include arcs.

Auxiliary View   An auxiliary view requires the selection of an existing drawing view edge to project around to display the true shape of a face.

Detail View   Detail is another type of view that is hard to make in a 2D system, especially if the geometry may change. For this view, you can use round or rectangular boundaries with smooth or rough edges for the view. The detail view can also be created with any scale.

> ▶
>
> A section view can be made to any depth or even defined outside a part to create a custom view.

Starting a New Drawing

Certification Objective

To have a view, you must first have a drawing. Beginning a new drawing is just like creating any other file in Inventor; you click a template and get to work.

1. On the Quick Access toolbar, click the New icon.

2. Make sure that 2014 Essentials.ipj is listed as the active project file near the bottom of the dialog box.

3. In the Create New File dialog box, select the Metric Template set on the left (see Figure 4.1).

4. Locate and double-click the ANSI (mm).idw template file.
   This opens a new drawing page in the Graphics window and updates the Ribbon to present the Place Views tab and its tools.

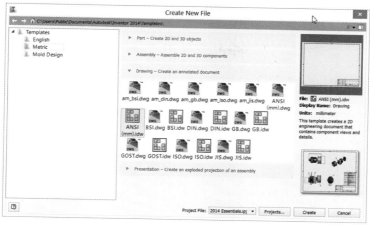

FIGURE 4.1 Templates are organized by file type.

Placing the Base and Projected Views

Now you can begin the process of creating your drawing. For this exercise, you will use the marking menu tools.

Certification
Objective

1. Right-click in the Graphics window, and move up to select the Base View tool, or right-click and drag up quickly to activate the tool.

 Base View...

2. In the Drawing View dialog box, click the icon to select an existing file at the end of the File drop-down list.

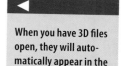

3. Using the same Open dialog box you used in Chapter 1, "Connecting to the Interface," use the parts shortcut in the upper left; then open the Parts\Chapter 04 folder and select the c04-01.ipt file.

   If you move your cursor over the drawing page, a preview appears, showing how the component will be sized and positioned on the page.

When you have 3D files open, they will automatically appear in the File drop-down list.

4. In the upper right of the Drawing View dialog box, click a few options in the Orientation field and see the effect each one has on the preview. Make sure you end up back at the Front view.

5. In the lower left of the dialog box, set the view scale to .7 and see the effect on the preview.

   At the bottom of many dialog boxes is an arrow icon. Clicking the arrow causes the dialog box to roll up to just its title bar. If a dialog box is in your way, use this option to make it easier to see your screen. Moving your cursor near the dialog box expands it to its full size, and reselecting the arrow holds the dialog box open.

6. Position the view similar to Figure 4.2, and click the mouse to place the drawing view.

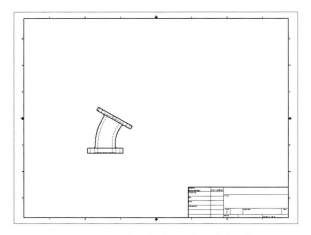

FIGURE 4.2 Placing the base view of the adapter

7. Move your cursor down and to the right and to the upper-right corner, and click the mouse to place the views that are shown in Figure 4.3. You must right-click and select Create from the context menu to generate the views.

The ability to quickly place the base view and the projected views in one step is a great way to rapidly see how the drawing will lay out.

TIP If you are creating drawings of very large or complex assemblies or have limited resources you can use the Raster View Only option in the Drawing View dialog box to speed up generation of the drawing views. This is useful to make sure the scaling and location of views is correct.

Once created, these views will keep a green border to indicate that they are raster views. The views can be updated to use higher-quality vector graphics once you are sure they will suit your needs.

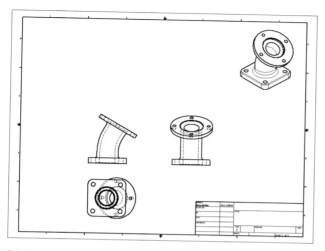

FIGURE 4.3 Locating the bottom, side, and isometric views

Placing a Section View

Section views are important for detailing machine parts and assemblies. In this exercise you will use the drawing views you have already created to define a section view:

1. Make certain that the 2014 Essentials project file is active; then open c04-02.idw from the Drawings\Chapter 04 folder.

2. Start the Section tool from the Create panel on the Place Views tab.

3. Click the projected view to the right of the base view to choose it as the parent of the section view.

4. As you move over and around the parent view, you will see dotted lines that show the alignment to geometry in the parent view. When an inference line appears from the center of the top face (see Figure 4.4), click to start creating a section line.

5. Drag the line down through the part to define the section line, and click a second point to start creating the section view (see Figure 4.5). Right-click and select Continue from the context menu to finish defining the section line.

6. When the Section View dialog box opens, you can change the scale and view identifier. Move your mouse to the right to see a preview of the view, as shown in Figure 4.6.

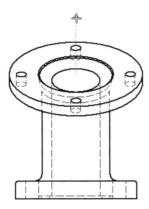

FIGURE 4.4 Inference lines will allow the section to be connected to geometry in the parent view.

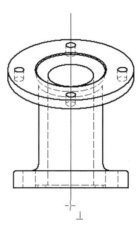

FIGURE 4.5 Include some overlap of the view in the section line length.

7. Click the location for the view to generate the view. Figure 4.7 shows the result.

It is possible to define a section view of nearly any existing drawing view. Auxiliary views are very similar to section views but are defined by model geometry.

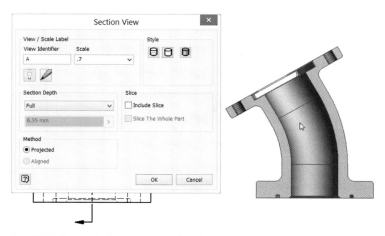

FIGURE 4.6 Placing the section view

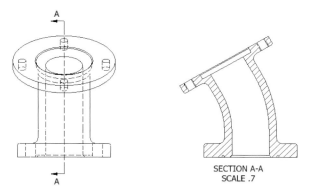

FIGURE 4.7 The finished section view

Creating an Auxiliary View

Auxiliary views can be a challenge to define accurately with a 2D CAD system. In this case, you want to more easily detail the round face of the part, which does not align with a standard orthographic projection.

1. Make certain that the 2014 Essentials project file is active, and then open c04-03.idw from the Drawings\Chapter 04 folder.

2. Move your cursor near the base view, and when it highlights, right-click to bring up the marking menu and context menu with other tools.

3. Select Create View ➢ Auxiliary View from the context menu.

4. After the Auxiliary View dialog box appears, click the angled edge on the top of the base view, as shown in Figure 4.8.

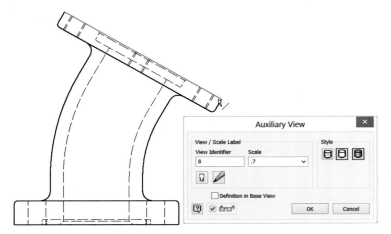

FIGURE 4.8   Selecting an edge for the projection reference

5. Click to place your view, as shown in Figure 4.9.

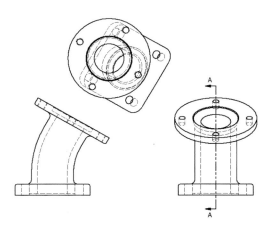

FIGURE 4.9   An easy auxiliary view

Creating a Detail View

The auxiliary view offers more clarity, but the section view offers an opportunity to see things otherwise obscured in an orthographic view. To do so, you'll create a detail view to add clarity:

1. Make certain that the 2014 Essentials project file is active, and then open c04-04.idw from the Drawings\Chapter 04 folder.

2. Start the Detail tool from the Ribbon or from the context menu.

3. Click the section view as the parent, which will open the Detail View dialog box.

4. Set Scale to 2 and Fence Shape to Rectangular; then click near the O-ring groove and drag the fence, as shown in Figure 4.10.

5. Drag the new view to any open spot on the page, and click to place it.

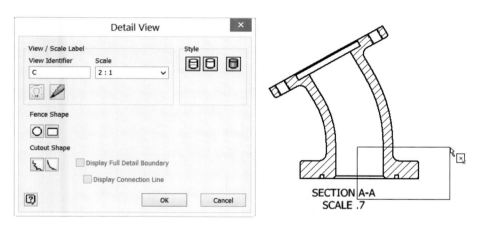

FIGURE 4.10 Setting up the focus of the detail view

Now that you have a collection of views, you should make the drawing look better as a whole. By moving, rotating, or even changing the appearance of the views, you can make a big difference.

Editing Views

The detail drawing is an important communication tool, so it is equally important to make it as clear as possible. In this section, you will focus on tools that will help you put the finishing touches on your drawings.

View Alignment

When views are created from other views, they inherit the appearance and scale of the parent view. Projected, auxiliary, and section views also inherit alignment from the parent view. Using this alignment and sometimes breaking the alignment make it simple to reorganize the drawing.

The simplest way to add clarity to a drawing is to shuffle the position of the drawing views. Doing this is a simple drag-and-drop operation.

1. Make certain that the 2014 Essentials project file is active, and then open c04-05.idw from the Drawings\Chapter 04 folder.

2. Click and drag the base view up and to the left. In Figure 4.11, you can see how the views projected from it and their children keep the alignment of orthographic projection.

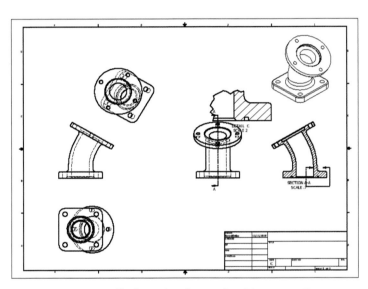

F I G U R E   4 . 1 1   Moving a view does not break its proper alignment with others.

3. Move the right-side view and the section views to give them more space. Notice that they remain in alignment with the base view but do not change its position.

4. Move the detail view to the left of the title block, as shown in Figure 4.12.

Moving views goes a long way toward making things look better. Sometimes, you might not want to maintain view alignment, or there may be times where it impedes the detailing of geometry.

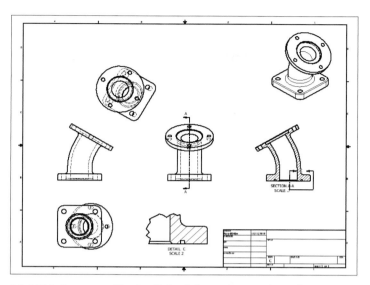

FIGURE 4.12 The detail view is free to be moved anywhere.

Changing Alignment

It is possible to edit a view and break its alignment to a parent view using simple right-click options or the tools in the Modify panel. You can also add alignment between views using tools from the same menus.

For example, to change the location of the auxiliary view, you need to break its alignment with the base view. To make it easier to dimension, you might want to rotate it as well. The Rotate tool will do both steps for you.

1. Make certain that the 2014 Essentials project file is active, and then open c04-06.idw from the Drawings\Chapter 04 folder.

2. Click the auxiliary view, right-click, and select Rotate from the context menu.

3. In the Rotate View dialog box, keep the rotation method By Edge, but set the alignment to Vertical and check that the direction is set to clockwise.

4. Click the straight edge on the right side of the view as shown in Figure 4.13.

5. Click OK after the part has been rotated.

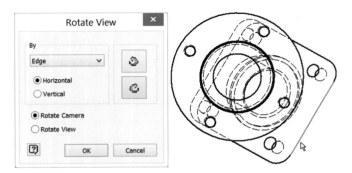

FIGURE 4.13   Rotating the drawing view can make detailing easier.

Now that the view has been rotated, you can detail it. But you might also want to reestablish an alignment with the base view to help others understand the geometry.

This exercise shows an uncommon and specialized drawing practice. It is built around the classic drafting idea of making it easier for people to recognize the geometry.

1. Make certain that the 2014 Essentials project file is active, and then open c04-07.idw from the Drawings\Chapter 04 folder.

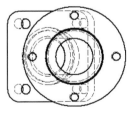

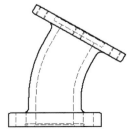

FIGURE 4.14
Views can be aligned to other views that were not originally their parent.

2. In the Modify panel, expand the options under Break Alignment and select Vertical.

3. Click the rotated auxiliary view and then the base view. See Figure 4.14 for the results.

Now that you think you have enough room between the views and they are aligned, you will make some of them easier to interpret.

View Appearance

Changing how a view looks or is scaled can have a dramatic effect on how easy it is to understand.

You can use a number of techniques, including removing hidden lines, adding shading, changing scale, or turning the visibility of selected entities off altogether. Any or a combination of these techniques can be done without changing the accuracy of the view.

A quick way to add clarity is to remove hidden lines when they're not needed to understand the geometry of the component. In this exercise, you'll also add shading.

1. Make certain that the 2014 Essentials project file is active, and then open c04-08.idw from the Drawings\Chapter 04 folder.

2. Double-click in the auxiliary view (upper left) to open the Drawing View dialog box.

3. Deselect the Style From Base check box.

4. To the left of the Style From Base icon, click the Hidden Line Removed icon, and click OK to close the dialog box and apply the change.

5. Double-click the isometric view to edit it.

6. Click the Shaded icon.

7. Select the Display Options tab in the dialog box, and then remove the check mark from Tangent Edges.

8. Change the scale of the view to .5.

9. Click OK to update the drawing. See Figure 4.15.

Removing unneeded hidden lines and adding shading can make the drawing easier to understand without adding more views. You could continue to make adjustments to the views for hours. Let's instead turn our attention to detailing the views you have.

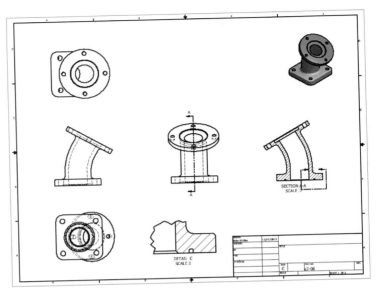

FIGURE 4.15   Updated drawing views

Adding Detail to Drawing Views

You can add many things to a drawing to make it complete. Much of the work is tedious but necessary. Reducing the time it takes to add all of the detail is one of the unsung strengths of Inventor. In the next several pages, you will see some great examples of the productivity tools for detailing and dimensioning.

Most of the tools you will be using are located on the Annotate tab. This is one time where Inventor doesn't automatically change the active tab, because everyone uses the detailing tools differently.

Center Marks and Centerlines

Certification Objective

You often need to add geometry to the drawing views to assist in locating dimensions. The next several exercises will walk you through the most commonly used tools for this purpose: center marks and centerlines.

Center marks make it easy to locate the center of a hole or radius.

1. Make certain that the 2014 Essentials project file is active, and then open c04-09.idw from the Drawings\Chapter 04 folder.

2. Click the Annotate tab in the Ribbon to change your active tab and see the Annotation tools.

3. Zoom in on the bottom view in the lower left of the drawing.

4. Click the Center Mark tool on the Symbols panel of the Annotate tab.

5. Move your cursor near the edge of one of the holes. When it highlights, click your mouse to place a center mark.

6. Repeat to place center marks on all four holes, as shown in Figure 4.16.

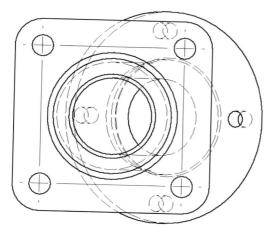

FIGURE 4.16 Placing center marks on holes by simply picking them

7. To connect the center marks, drag the center mark extensions toward each other.

Now that you've added and edited center marks for a basic rectangular hole pattern, you can move on to bolt circles.

Center marks and centerlines have many uses in Inventor. The ability to easily make and drag-edit the extents of these marks helps you save time cleaning up the drawing. They can also be used to show linear alignments between holes or even bolt circles.

You can also easily create a series of centerlines in an arc based on a center.

▶

1. Make certain that the 2014 Essentials project file is active, and then open c04-10.idw from the Drawings\Chapter 04 folder.

2. Zoom in on the rotated auxiliary view in the upper-left corner.

3. Select the Centerline tool from the Symbols panel of the Annotate tab.

4. Begin by selecting the hole at 12 o'clock of the round face; then click the holes at 3 o'clock, 6 o'clock, and 9 o'clock, as shown in Figure 4.17.

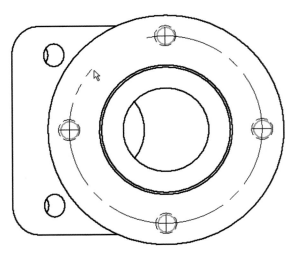

FIGURE 4.17 As you place the center marks for the bolt circle, the centerline previews.

5. Finish the Bolt circle by clicking the 12 o'clock hole again; then right-click and select Create from the context menu. See Figure 4.18.

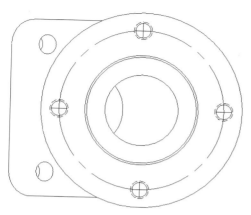

FIGURE 4.18 Generating a bolt circle with the Centerline tool

6. Press the Esc key to leave the Centerline tool.

You can use the Centerline Bisector tool on straight or curved segments to find the midline between them. Straight segments do not need to be parallel.

1. Make certain that the 2014 Essentials project file is active, and then open c04-11.idw from the Drawings\Chapter 04 folder.

2. Zoom in on the section view on the right side of the drawing.

3. Select the Centerline Bisector tool on the Symbols panel of the Annotate tab.

4. Click the Orange lines in the lower left and lower right of the open portion of the part to place a straight segment.

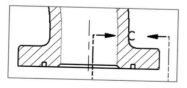

5. Click the green arcs on the left and right to place a curved segment.

6. Place another straight segment by picking the blue angled line at the top left and top right.

You can also extend the straight segments beyond their original size for clarity. See Figure 4.19.

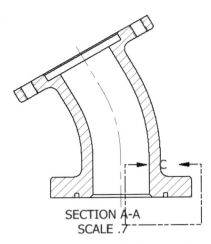

FIGURE 4.19 Placing a complex centerline bisector

While you are looking at the section view, you should improve its appearance.

Editing a Detail View Placement and Callout

Previously, you moved views around to improve the layout of the drawing. Callouts, section lines, and detail boundaries can also be edited to improve their clarity.

1. Make certain that the 2014 Essentials project file is active, and then open c04-12.idw from the Drawings\Chapter 04 folder.

2. Zoom in on the section view.

3. Select the letter C in the detail view boundary, and drag it to where the upper-right corner is now.
   In addition to relocating the callout, you can change the letter by editing the text.

4. Click the boundary. When the corners and center highlight, drag the center upward and to the right a short distance.

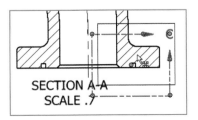

5. Select the *SECTION A-A* text, and drag it below the view so it is legible. Figure 4.20 shows the result.

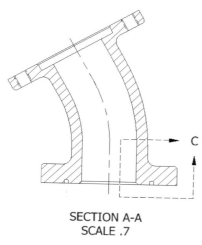

SECTION A-A
SCALE .7

FIGURE 4.20 Relocate the section view label for a better drawing view.

Now that the views have been updated with new or repositioned elements, you can turn your attention to dimensioning.

Dimensioning

If you've used traditional 2D CAD tools in the past for drafting, then the tools Inventor uses to apply dimensions should seem familiar but different at the same time. Rather than offer the user a broad array of dimensioning tools, Inventor uses intelligence that changes the type of dimension based on the selected geometry. Add the ability to change many options while you are placing the dimension, and you are sure to appreciate the power available to you.

The General Dimension Tool

Certification
Objective

To place individual dimensions in the drawing, the General Dimension tool will give you almost anything you will need. For horizontal, vertical, aligned, radial, and diameter dimensions, you can just use this tool.

1. Make certain that the 2014 Essentials project file is active, and then open c04-13.idw from the Drawings\Chapter 04 folder.

2. Zoom in on the bottom view in the lower left of the drawing.

Dimension

3. From the Dimension panel of the Annotate tab, select the Dimension tool.

4. Click the center marks in the lower- and upper-right corners (Figure 4.21).

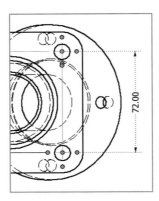

F I G U R E 4 . 2 1 Click the center marks or other geometry to place dimensions.

5. Move your cursor to the right. You will see the dimension change to dotted lines when you have good spacing away from the geometry.

6. Click to place the dimension.

7. When the Edit Dimension dialog box appears, deselect the Edit Dimension When Created check box, and click OK to close the dialog box.

8. The General Dimension tool is still running. Click the radius in the lower-right corner, and notice that you are now creating a radial dimension. This dimension will snap at common angles.

9. Click to place the new dimension (Figure 4.22).

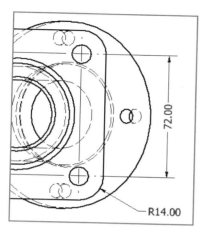

FIGURE 4.22 Adding a radial dimension

10. Pan up to the base view.

11. Click the upper edge of the base of the part and then the lower edge of the angle face. Notice that the dimension now turns to an angular dimension.

12. Place the new dimension to the side, as shown in Figure 4.23.

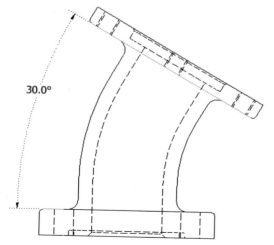

FIGURE 4.23 Even angular dimensions can be placed with the General Dimension tool.

13. Select the top edge of the angle portion.

14. Place the linear diameter above the drawing view (Figure 4.24).

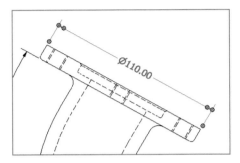

FIGURE 4.24 The geometry selected determines the type of dimension placed.

As you see, the General Dimension tool is incredibly flexible.

The Baseline and Baseline Set Dimension Tools

When you need to place several dimensions from the same datum, you can use the baseline dimensioning tools, Baseline and Baseline Set. The difference between the two is that Baseline Set keeps the placed dimensions' styles and precision linked along with other variables.

1. Make certain that the 2014 Essentials project file is active, and then open c04-14.idw from the Drawings\Chapter 04 folder.

2. Zoom in on the bottom view in the lower left of the drawing.

3. From the Dimension panel of the Annotate tab, expand the Baseline tool, and select Baseline Set.
   The tool starts immediately and waits for you to select the geometry.

4. Click the left edge of the square face, the top-left hole, the top-right hole, and the right edge.

5. Right-click and select Continue from the context menu to stop selecting geometry and begin placing dimensions.

6. When the preview of the dimensions appears, move them above the view.

7. Click to place the dimensions, and then right-click and select Create from the context menu to finish. See Figure 4.25.

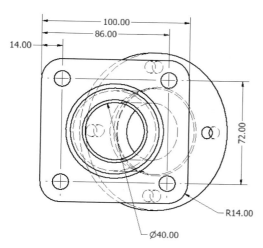

FIGURE 4.25 Placing multiple dimensions with the Baseline Dimension tool

The Chain and Chain Set Dimension Tools

Used commonly in architecture and when part cost is more important than high precision, the Chain dimensioning tools work very much like the Baseline tools but place dimensions in a series rather than from a common datum.

1. Make certain that the 2014 Essentials project file is active, and then open c04-15.idw from the Drawings\Chapter 04 folder.

2. Zoom in on the bottom view in the lower left of the drawing.

3. From the Dimension panel of the Annotate tab, select the Chain Dimension tool.

   The tool starts immediately and waits for you to select the geometry.

4. Click the top edge of the square face, the top-left hole, the bottom-left hole, and the bottom edge.

5. Right-click and select Continue from the context menu to stop selecting geometry and begin placing dimensions.

6. The preview of the dimensions will appear. Move them to the left of the view.

7. Click to place the dimensions; then right-click and select Create from the context menu to finish. See Figure 4.26.

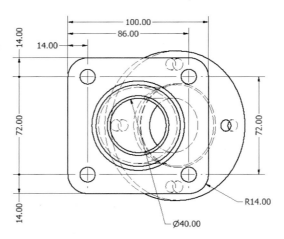

FIGURE 4.26 Chain dimensioning is much easier with a specific tool.

The Ordinate and Ordinate Set Dimension Tools

The Ordinate dimensioning tools work much like the Baseline tools but place the dimension value at the end of the extension line, as shown in Figure 4.27.

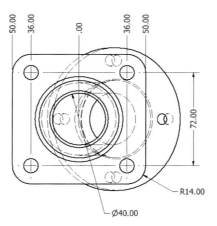

FIGURE 4.27 Ordinate dimensions work with the same workflow as chain dimensions.

The difference between the two is that the Ordinate Dimension tool prompts for the placement of a datum point in the view. The Ordinate Set dimension tool does not, and it therefore has a workflow that is essentially the same as that of the Baseline Dimension tool.

Editing Dimensions

When you placed the baseline dimension set, it was placed almost as a unit. The dimensions share many properties. You can still change these properties, and if need be, you can make a dimension independent.

1. Make certain that the 2014 Essentials project file is active, and then open c04-16.idw from the Drawings\Chapter 04 folder.

2. Zoom in on the bottom view in the lower left of the drawing.

3. Click the 72.00 dimension on the left.
   Once you've selected the dimension, two drop-downs will become active in the Format panel on the Annotate tab.

4. On the lower drop-down, change the Dimension style to Default–mm [in] (ANSI).

5. Right-click the 14.00 dimension in the baseline set, and select Detach Member from the context menu.

6. Double-click the 14.00 dimension.

7. When the Edit Dimension dialog box opens, select the Precision And Tolerance tab.

8. In the Precision field on the right, use the Primary Unit drop-down to change the value to 1.1. Click OK. See Figure 4.28.

You can also delete a member and use the Arrange tool to bring the dimensions back in order.

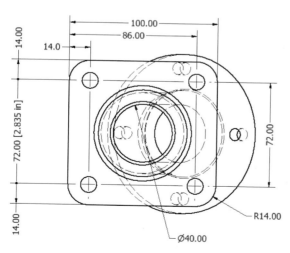

FIGURE 4.28 The dimensions can still be edited after being placed.

The Edit Dimension dialog box offers a lot of options that I didn't cover in the exercise, including tolerancing, fits, and even the ability to define an inspection dimension.

The Hole and Thread Notes Tool

  The part modeling tools give you the ability to define holes by thread and their clearances for fasteners based on standards. To document these features, use the Hole and Thread note tool, which extracts the information directly from the feature.

1. Make certain that the 2014 Essentials project file is active, and then open c04-17.idw from the Drawings\Chapter 04 folder.

2. Zoom in on the rotated auxiliary view in the upper left of the drawing.

3. Find and start the Hole and Thread tool in the Feature Notes panel of the Annotate tab.

4. Click the hole at 3 o'clock, and place the dimension above and to the right of the hole.

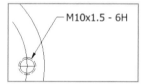

5. Press the Esc key to end the tool.
   Now you can edit the annotation to add detail.

6. Double-click the Hole note.

7. In the Edit Hole Note dialog box that appears, press the Home key or click the cursor before the text <THDCD>.

 8. Click the Quantity Note icon to add the ability to display the number of like holes in the hole note (see Figure 4.29). Then click OK. See Figure 4.30 to see the completed hole note.

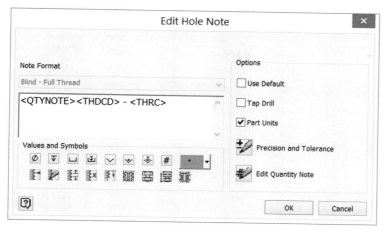

FIGURE 4.29   The Edit Hole Note dialog box

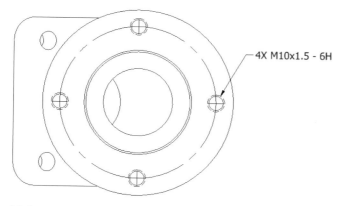

4X M10x1.5 - 6H

FIGURE 4.30   The Hole note can also display the number of holes.

Retrieving Model Dimensions

The parametric dimensions that are used in sketches to construct the parts can also be used to detail the part.

1. Make certain that the 2014 Essentials project file is active, and then open c04-18.idw from the Drawings\Chapter 04 folder.

2. Zoom in on the rotated auxiliary view in the upper left of the drawing.

3. Select the Customize tool from the Options panel on the Tools tab.

4. Select Full Menu from the Overflow Menu options drop-down in the lower left of the Marking Menu tab in the Customize dialog box, and then click Close.

   This will give you an expanded set of tools to use.

5. Right-click in the drawing view, and select Retrieve Dimensions from the context menu, or select the Retrieve tool from the Dimension panel of the Annotate tab.

6. Once the Retrieve Dimensions dialog box is open, select one of the tapped holes in the drawing view.

   The drill diameter of the holes and the diameter of the bolt circle will be displayed. When Inventor needs more information from the user, an icon will appear in the dialog box with a red arrow. Often, the button will be depressed, and selection can be made immediately. Sometimes, you will need to select the button to begin picking entities.

7. Click the icon to select dimensions, and click the 90 mm diameter of the bolt circle.

8. Click OK to close the dialog box and include the dimension in the drawing view, as shown in Figure 4.31.

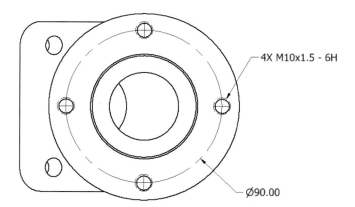

FIGURE 4.31 Reusing model dimensions in the drawing view

TIP You might want to restore your Overflow Menu setting to the default. The short menu should provide you with the tools that you would normally use.

Associativity

The ability to create drawing views and annotate them should be clear by this point. But I haven't demonstrated the advantage of creating 2D parts from 3D geometry. When you edit a 3D part, any change is reflected in the 2D drawing.

This capability is key for productivity and demonstrates the final value of creating 2D drawings from 3D parts. The technique used in this exercise is not the most common way of making the changes, but it does show a technique that is available.

1. Make certain that the 2014 Essentials project file is active, and then open c04-19.idw from the Drawings\Chapter 04 folder.

2. Zoom in on the rotated auxiliary view in the upper left of the drawing.

3. Right-click the 90 mm diameter dimension. Because it is a model dimension, it will offer a special option.

4. Select Edit Model Dimension from the context menu.

5. In the Edit Dimension dialog box, change the value to 85 mm, and click the green check mark to make the change. See Figure 4.32 for the results.

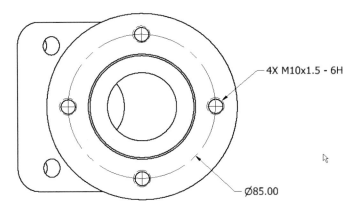

FIGURE 4.32 Any change to the geometry of the model is reflected in the drawing views.

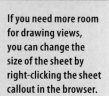

If you need more room for drawing views, you can change the size of the sheet by right-clicking the sheet callout in the browser.

The drawing view that you were focused on is not the only view that was changed. All views of this part on all pages in any drawing file will be updated based on this change.

Replace Model Reference

Many parts are variations on an existing design. Oftentimes, you will create a whole new part number for a minor difference. Inventor can copy an existing drawing and replace the source of its drawing views. In the following exercise, you will use the finished drawing for the c04-01.ipt file as the source for a new drawing based on a new part.

1. Make certain that the 2014 Essentials project file is active, and then open c04-20.idw from the Drawings\Chapter 04 folder.

2. Start the Replace Model Reference tool from the Modify panel on the Manage tab.

3. In the Replace Model Reference dialog box, select the current drawing, and then click the Browse icon. In this case, there is only one model referenced.

4. Browse, select the Parts\Chapter 04\c04-03.ipt file, and click Open.

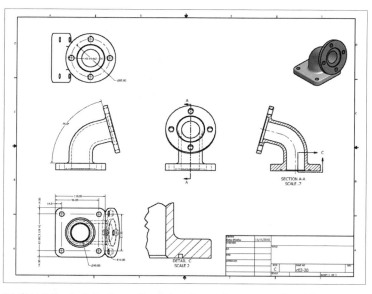

FIGURE 4.33 Some dimension cleanup is necessary, but this was finished in seconds.

5. When the warning that you are changing the model appears, click Yes; then click OK to update the drawing and see the change, as shown in Figure 4.33.

Some annotation might be lost or have to be relocated when making a change like this. For parts that have the same geometry but different sizes, this technique can offer a huge time savings for detailing the new part.

THE ESSENTIALS AND BEYOND

There are countless ways to approach creating your detail drawings, and Inventor maintains Autodesk's legacy of flexibility but adds a level of simplicity that's not possible when creating 2D drawings from scratch.

ADDITIONAL EXERCISES

▶ Experiment with all the tools on the Place Views tab.

▶ Use DWG templates to create drawings that can be read using AutoCAD.

▶ Work with individual dimensions and dimension sets to see what method you might typically want to use.

▶ Try breaking the alignment of the section view to see that it still maintains its definition.

CHAPTER 5

Introducing Assembly Modeling

Few people need to work on parts that exist on their own. Not only is working with assemblies necessary, but in many ways, Autodesk® Inventor® software works best in the assembly environment. To become proficient with the Inventor assembly environment, you must understand how to use assembly constraints. Once you're familiar with the tools, it is best to think about how things fit together from a more elemental standpoint.

▶ **Creating an assembly**

▶ **Understanding grounded components**

▶ **Applying assembly constraints**

▶ **Working with the Content Center**

▶ **Using the Bolted Connection Generator**

▶ **Joints: An Alternative to Constraints**

Creating an Assembly

Certification Objective

To create an assembly, you follow the same basic process as beginning any other type of file in Inventor. Starting a new assembly file changes the Ribbon to display tools that you need for the assembly process.

1. Open the New dialog box, navigate to the Metric template folder, and start a new assembly using the Standard (mm).iam template.

2. Select Place Component from the marking menu or in the Component panel of the Assemble tab.

3. Go to the Parts folder and then to Chapter 05; double-click the c05-01.ipt file and click OK.

4. A preview of the part will appear, and you are able to position it in the assembly. Right-click and select Place Grounded At Origin from the marking menu.

5. Press the Esc key to finish placing components.

6. Orbit your assembly so that you can see the Top, Front, and Right faces of the ViewCube®.

7. Begin placing another component in the assembly, this time selecting the c05-02.ipt file from the same folder.

8. When the preview appears, right-click and select Rotate Z 90° from the marking menu; then click Rotate Y 90° two times to reorient the part.

9. Click in the Design window to the right of the first part to create one instance and then end the command (see Figure 5.1 for reference).

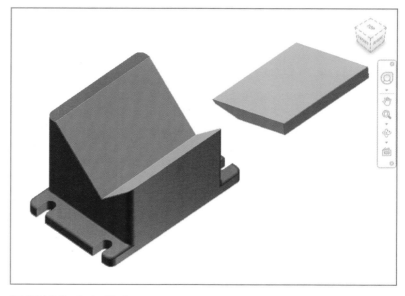

FIGURE 5.1   Placing components into a new assembly

10. Set this point of view to be your new Home view.

Take a moment to look at your Browser. The components you placed appear just as if you'd placed views in a drawing or features in a part. You will see the name of the part displayed, followed by a colon and an additional number. This number denotes which instance of the component you are looking at.

Understanding Grounded Components

The first component placed in an assembly is referred to as the *base component* and, as in the real world, should be the component that the largest number of other components are attached to or rely on for their foundation.

In the Browser, notice a thumbtack icon overlaying the cube icon that represents the part. This thumbtack means the component is *grounded*, which means you don't need to constrain the base part to the coordinate system of the assembly file. Any or all components can be grounded, and the base component can be "ungrounded." You can place a base component without grounding it if you want to be able to reorient it in the assembly after it has been placed.

Applying Assembly Constraints

The purpose of constraining components together in an assembly is to mimic the behavior of components in the real world. The constraints you apply do this by removing degrees of freedom from the components. Although it is not necessary to remove all the degrees of freedom, you will need to remove as many as needed to properly position the part. Inventor has relatively few constraint tools, but some of them can be used in many ways.

The Place Constraint dialog box has four tabs: Assembly, Motion, Transitional, and Constraint Set. Each accesses tools that offer different ways to solve your assembly. The selections area on each tab and with each type offers buttons with different colors that are reflected in the assembly as you select the entities. In this section, you'll use exercises to better understand the most commonly used constraint tools.

The Assembly tab (Figure 5.2) offers five types of assembly constraint: Mate, Insert, Angular, Tangent, and Symmetry. Most of these constraints offer more than one way of defining them called a solution. In addition, the Transitional tab offers its own constraint. For the most part, you will see the options each constraint offers by looking at the icons for its solutions.

If you know you need three instances of a component, you can also drag an instance from the Browser into the Design window to place another.

Certification Objective

   c05-01:1

The Browser display of a component will show the component type (part or assembly), the grounded status, the part name, and after the colon, the instance of the component, which is a number that is added sequentially and automatically.

Certification Objective

FIGURE 5.2 The Assembly tab of the Place Constraint dialog box

The Mate Constraint

This tool has two primary options: a Mate solution and a Flush solution. Each will reposition components based on the selected geometry with the ability to apply an Offset value in the dialog box that will create a gap or an interference between the components, depending on whether the value is positive or negative and how the parts are aligned.

The Mate solution works on planes, axes, and points, and it behaves as though it puts these entities in opposition to one another. If you put a box on top of another box in the real world, you could say that the top of the lower box was pushing against the bottom of the upper one. In Inventor, this would be a Mate/Mate constraint.

First you will use the Mate solution in your assembly:

1. Make certain that the 2014 Essentials project file is active and then open c05-01.iam from the Assemblies\Chapter 05 folder.

2. Launch the Constrain tool from the Relationships panel or from the marking menu.

3. Set Constraint Type to Mate, and leave Solution at its default (Mate).

4. In the Selection group, the button for 1 will be depressed. Click the blue face on the purple part for the first selection.

5. Once the first object is clicked, the second one will be enabled. Click the blue face on the red part.

   The two parts automatically snap together (see Figure 5.3). You can turn this off by deselecting the Preview icon in the dialog box if you have a system with limited performance.

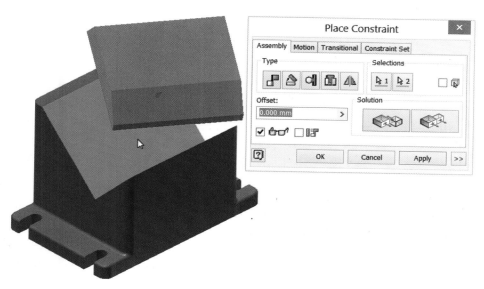

FIGURE 5.3 The components move into position as you apply constraints.

6. Click Apply in the dialog box or right-click and select Apply from the context menu to create the first constraint and prepare for another.

7. Orbit your assembly, and select the yellow face of the red part and the yellow face of the purple part.

8. Click OK to place the constraint and close the tool.

9. Drag the purple part in the assembly. You can see that it is still able to slide from side to side.

The Flush solution works only on planes and controls the alignment of faces to one another.

Similar to a Mate constraint with a Mate solution, the Symmetry constraint can keep two components spaced equidistant from a plane or planar face.

1. Verify that the 2014 Essentials project file is active and then open the c05-02.iam file from the Assemblies\Chapter 05 folder.

2. Launch the Constrain tool from the Relationships panel or the marking menu.

3. Set Constraint Type to Mate, and set Solution to Flush.

4. Select the sides of the two components, as shown in Figure 5.4.

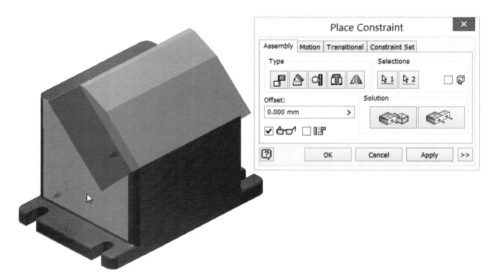

FIGURE 5.4   The Flush constraint aligns faces.

5. With Preview on, the components will align. Click OK to create the constraint.

Certification Objective

The simple Mate constraint and its solutions cover the needs of many Inventor users. To better understand which constraints are missing in an assembly, you can use the Degrees Of Freedom (DOF) tool.

DEGREES OF FREEDOM

As mentioned earlier in this chapter, Inventor uses the concept of degrees of freedom to describe what movement can be realized by a component. Unless a component is grounded or constrained, it has six degrees of freedom—three rotational and three translational around and along the x-, y-, and z-axes. As you apply constraints, the ability to move in particular directions is removed.

The Insert Constraint

The Insert constraint is a hybrid and combines a Mate constraint on the axis and between the adjacent faces of curved edges. It has many uses, but it is most commonly used to put round parts in round holes.

1. Verify that the 2014 Essentials project file is active and then open the c05-03.iam file from the Assemblies\Chapter 05 folder.

2. Switch the Ribbon to the View tab.

3. Click the Degrees Of Freedom tool in the Visibility panel.
   An icon appears on the purple part, showing that it has all six degrees of freedom.

4. Launch the Constrain tool from the Assemble tab in the Relationships panel or the marking menu.

5. Set Constraint Type to Insert, and leave Solution set to Opposed.

6. Select the circular edge of the hole on the purple part's flat face, and then select the circular edge of the hole on the red part, as shown in Figure 5.5.

7. When the parts are aligned, click OK.
   The Degrees Of Freedom icon changes to show one remaining rotation of freedom.

8. Drag the purple part to see how it moves.

◀

Editing an assembly constraint will give you access to setting a range for the values of the constraint rather than it being a static number. This allows for easy range of motion studies.

Being able to exercise the remaining degrees of freedom is not limited to a single component. You can use this to show how a mechanism works.

FIGURE 5.5 Applying constraints does not automatically restrict all movement.

The Angular Constraint

There are three solutions for the Angular constraint, allowing the flexibility to, for example, set the axis of rotation apart from the components, along with other options that give less control. For this exercise, you will use the most precise solution, the Explicit Reference Vector tool.

1. Verify that the 2014 Essentials project file is active, and then open the c05-04.iam file from the Assemblies\Chapter 05 folder.

2. Launch the Constrain tool from the Assemble tab in the Relationships panel or the marking menu.

3. Set Constraint Type to Angle, and make sure the solution type is set to Explicit Reference Vector.

4. For the first selection, click the narrow face on the side of the purple part.

5. For the second, click the face on the red part, as shown in Figure 5.6.

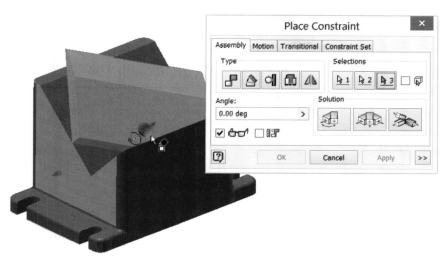

FIGURE 5.6 Complexity in a constraint also adds reliability and stability.

6. For the third, click the cylindrical face of the hole in the purple part.

7. Click OK to complete constraining of the purple part. The purple part can no longer rotate freely around the axis of the hole. This constraint is a great option for aligning faces that aren't flush on all sides.

The Tangent Constraint

When rounded faces need to stay in contact with other rounded faces, the Tangent constraint is often the only tool you will need. The constraint can have Inside or Outside solutions.

1. Verify that the 2014 Essentials project file is active and then open the c05-05.iam file from the Assemblies\Chapter 05 folder.

2. Launch the Constrain tool from the Assemble tab in the Relationships panel or the marking menu.

3. Set Constraint Type to Tangent, and make sure the solution type is set to Outside.

4. Make your first selection, the blue, curved face on the orange part.

5. Make the second selection, the blue face on the gray part.

A good use for an Inside solution would be a pin following a curved face. An Outside solution would be rollers in contact with each other.

6. Click OK to bring the parts together.

7. Drag the orange part to see that the selected faces stay in contact.

The tangency added follows the entire cylindrical face. If the face isn't a full cylinder, components constrained to it will follow the curvature of the face even if it moves off the end of the face.

The Transitional Constraint

A fifth type of constraint is available on its own tab. Where the Assembly tab's Tangent constraint works well for simple curved faces, the Transitional constraint works best with changing faces such as cams.

1. Verify that the 2014 Essentials project file is active, and then open the c05-06.iam file from the Assemblies\Chapter 05 folder.

2. Launch the Constrain tool from the Assemble tab in the Relationships panel or the marking menu.

3. In the Place Constraint dialog box, switch the tab to Transitional.

4. Click the yellow face on the orange part for the first selection.

5. Click the yellow face on the gray part for the second selection.

6. Click OK to accept the constraint. See Figure 5.7.

> The Motion constraint class is explored in Chapter 9, "Advanced Assembly and Engineering Tools."

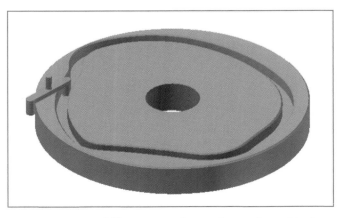

F I G U R E   5 . 7   Different types of constraints can be combined to create useful models.

7. Drag the orange part to see how the Transitional and Tangent constraints work together to position the two components.

Adding a Transitional constraint offers more flexibility on the type of faces to which you can constrain than the Tangent constraint. Now, you will take a look at working with other data sources in the assembly.

Working with the Content Center

During installation, you have the option of installing various types of standard components. Depending on the options you choose, you will have tens or hundreds of thousands of standard components available to you. Rather than creating part files for bolts, nuts, and so on, you can select them from a library.

Certification Objective

1. Verify that the 2014 Essentials project file is active, and then open the c05-07.iam file from the Assemblies\Chapter 05 folder.

2. Locate the Place From Content Center tool under the Place tool in the Component panel of the Assemble tab.

   This will launch the Place From Content Center dialog box (Figure 5.8), where you can locate many kinds of standard components.

Place from Content Center

Many companies choose to limit the number of components by copying the ones they use to a custom library and limiting access to others.

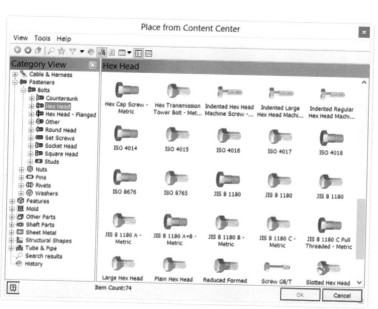

FIGURE 5.8 Hundreds of thousands of standard components are built into Inventor.

3. Expand the categories by selecting Fasteners ≻ Bolts ≻ Hex Head.

4. Double-click ISO 4017.

   When the dialog box disappears, a small preview of the fastener will appear. This preview is one of the available sizes for the bolt.

5. Hover over the edge of the hole in the purple part until the fastener resizes based on the hole's size.

6. After the bolt resizes, click when the edge of the hole highlights in red (Figure 5.9).

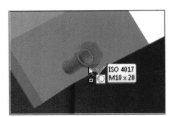

FIGURE 5.9 The fastener will resize based on an existing hole.

This temporarily places a detailed preview the bolt in the hole.

7. Without ending the Place Constraint tool, change the Ribbon to the View tab, and change Visual Style to Shaded With Hidden Edges.

8. Click the Front face of the ViewCube to change the view.

9. Click and drag the red arrow at the tip of the bolt until it shows that the fastener size is at M10 35, and then release to set the length (Figure 5.10).

FIGURE 5.10 Fastener lengths can be changed in standard sizes.

10. Click the check mark in the AutoDrop mini-dialog box to place the fastener in the assembly.

11. The tool will place the fastener and then offer to place another bolt. Press Esc to finish the command.

12. Switch to the Home view, and move the fastener to see that it is constrained to the hole.

    The remaining degree of freedom on the fastener does not need to be removed. As long as the fastener is seated and aligned with the hole, it is effective in the assembly.

    Even when a component is fully constrained it can be moved or rotated for additional clarity using the Free Move or Free Rotate tool.

13. Select the Free Move tool in the Position panel of the Assemble tab.   Free Move

14. Click and drag the purple part into an open part of the graphic window and release the mouse button.

    When you release the mouse, glyphs will appear to show the position of assembly constraints that are related to the part, as shown in Figure 5.11

FIGURE 5.11 Free Move will help you see what constraints are placed.

15. Click OK on the marking menu to leave the purple part in this position until the assembly is updated.

    You can pan, zoom, or rotate and even apply additional constraints to the displaced part, which is very useful in a large assembly or when you have complex parts to select features from.

16. Click the Update icon in the title bar to restore the assembly constraints that you applied.

    A completed assembly will hold together similar to the assembly in the real world. That doesn't mean you are limited to only things you can do in the real world, so you must be diligent and think about what realities exist for creating your designs.

Using the Bolted Connection Generator

The process of applying holes to components and then positioning fasteners is not difficult, but it does put the focus on the features of the software rather than the needs of the design.

The Bolted Connection Generator tool will allow you to put standard components into an assembly and use engineering calculations to be sure you are using the correct component.

1. Verify that the 2014 Essentials project file is active, and then open the c05-08.iam file from the Assemblies\Chapter 05 folder.

2. Activate the Design tab on the Ribbon.

3. Start the Bolted Connection tool.

    The Bolted Connection Component Generator dialog box (Figure 5.12) has a number of options, but it works very much like the Hole feature in a part model.

4. In the Type group, set hole type to a Blind connection Type.

5. For the start plane, use the yellow face of the purple part.

6. For Linear Edge 1, click the left edge of the yellow face. If you are prompted for a value for this dimension, use **50 mm**, as shown in Figure 5.13.

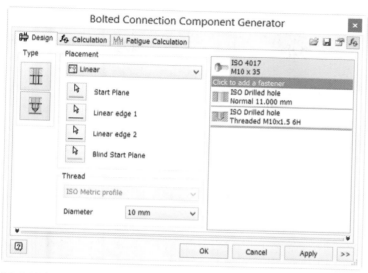

FIGURE 5.12 The Bolted Connection Component Generator dialog box

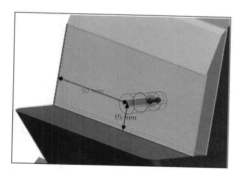

FIGURE 5.13 The dimensions locate
the placement of the bolted connection.

7. Click the bottom edge for Linear Edge 2. If you are prompted for a value for this dimension, use **15 mm** for the value.

8. Click the face where the two parts are in contact to select the blind start plane. You will be able to pick through the purple part.

9. In the Thread group, set thread class to ISO Metric Profile.

10. Set the diameter to **10 mm**.

11. Click where it says *Click to add a fastener* in the column on the right.

After a short time, a list of bolts that will fit the hole as currently defined appears. You can filter the list to see only those bolts of a single standard.

12. In the component selection pop-up, change Standard to ISO to filter the options. See Figure 5.14.

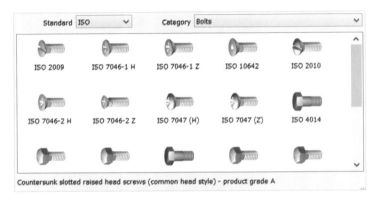

FIGURE 5.14 Filtering fasteners based on standards makes selection easier.

13. Click ISO 4017.

14. Change the view in the Design window to the Front view of the assembly.

15. Drag the red arrow at the end of the bolt until it shows you are placing an M10 35 bolt (Figure 5.15).

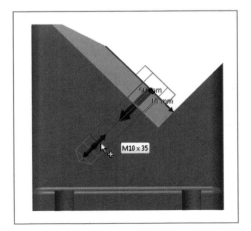

FIGURE 5.15 In a bolted connection, you can drag the hole depth and bolt length.

16. Click OK to finish the tool and click OK again to approve the creation of a file.

17. Double-click the red part to see that a tapped hole has been added to it.

18. Click the Return tool at the end of the Ribbon to return to the assembly.

Placing fasteners using this tool saves time over placing holes in advance and using the Content Center. If the assembly changes, you might have to edit the bolted connection and click OK to update it. I will review Design Accelerator tools in Chapter 9.

Joints: An Alternative to Constraints

If you have Inventor Professional and want to do dynamic simulation, you can save yourself a few steps by applying joints to the assembly rather than con-straints. If you don't own Inventor Professional, you might still find joints a nice alternative. You can often use them to combine what would be more than one constraint into a single joint.

1. Verify that the 2014 Essentials project file is active, and then open c05-09.iam from the Assemblies\Chapter 05 folder.

2. Start the Joint tool from the Relationships panel of the Assemble tab.

Joint

3. Click the edge of the hole on the blue face of the purple part and then the edge of the hole on the blue face of the red part.
   After you select the geometry, an animation plays that highlights any remaining degrees of freedom.
   Based on the geometry you select, the Joint tool will select the type of joint it thinks should be applied. You can override this if there is more to the relationship than Inventor anticipates. In this case it will assume you want a Rotational joint, but it doesn't understand that the bolt can translate in and out of the hole, so you will need to change the type of joint that will be applied.

4. Use the pull-down menu in the mini-toolbar to change the joint to a Cylindrical joint, as shown in Figure 5.16.

FIGURE 5.16 Joint types can be changed through a pull-down.

This will allow the purple part to slide along the axis of the holes, but you also need to restrict the rotation of the part as you did with an angle constraint earlier in this chapter. To do this, you can simply add an align relationship.

5. Pick the icon for first alignment in the mini-toolbar and then select the yellow face on the purple part (if it's not already highlighted).

6. Click the icon for the second alignment and then pick the yellow face of the red part.

7. Click the Invert Alignment icon if needed to position the part properly; then right-click and select Apply from the marking menu.

8. Pick a round edge on the bottom of the head of the bolt as you did for the insert constraint previously.

9. Pick the edge of the hole in the purple part. You may need to click Flip Component.

10. Change the Rotational joint to a Rigid joint to remove all degrees of freedom from it; then click the green check mark to approve the creation of the joint.

11. Expand the C05-10 component in the Browser. Right-click on the Cylindrical joint and choose Edit from the context menu.

12. If the full Edit Joint dialog box doesn't appear onscreen, click the Show/Close dialog toggle in the mini-toolbar.

13. In the Edit Joint dialog box, select the Limits tab.

14. Click Start and End under Angular, and then set Start to **0 deg** and End to **.001 deg**.

    The angle needs a real value to work, and as in the real world, nothing is absolutely perfect.

15. Click Start and End under Linear and set their values to **0 mm** and **−30 mm**, respectively. The dialog box should look like Figure 5.17.

FIGURE 5.17 Adding limits to the Cylindrical joint

16. Click OK to close the dialog box.

17. Click and drag on the purple part or bolt to see that you have restricted movement in the assembly.

18. Right-click the Rigid joint, pick edit from the context menu and change it to `Rotational.

19. Click OK to complete the change and move the assembly to see that the bolt now has a rotational degree of freedom.

The Joint tool is new to Inventor and it remains to be seen if it will ever replace the Constrain tools. The ability to combine what would be many constraints makes the Joint tool an appealing option.

THE ESSENTIALS AND BEYOND

This chapter covered the basics of the assembly process of Inventor. Because of the simplicity of assemblies in Inventor, you learned about the most commonly used tools and even some more advanced options. Assemblies represent the completion of a design and must be able to capture its function. Using Assembly or Joint options to help others see how your product will work give it a better ability to communicate the function of your design.

ADDITIONAL EXERCISES

▶ Review the components in the Content Center to see the extent of the options.

▶ Experiment with the Joint tool for doing all of the exercises.

▶ Work with different fasteners and standards in the Bolted Connection Component Generator dialog box.

CHAPTER 6

Exploring Part Modeling

In Chapter 3, "Introducing Part Modeling," you used a number of key features in the Autodesk® Inventor® 2014 program that are most commonly used in 3D solid modeling. You used these tools in the context of creating a component to see how these tools add features to, and remove features from, a part.

In this chapter you will take a step back from this workflow-oriented process and focus on the elements and options of powerful tools. You will work with techniques that go beyond the fundamentals. You will also gain a better understanding of where the power of parametric modeling really lies: in the editing of parametric models.

▶ **Modeling using Primitives**

▶ **Understanding work features**

▶ **Using sketched feature termination options**

▶ **Editing and redefining work features**

Modeling using Primitives

Primitives can be used as a shortcut to define a solid model, but once you're familiar with them you may find yourself using them to quickly "sketch" concepts or create a quick mockup of a component to explain an idea to others. You can use a handful of primitive shapes to build a complex model in no time.

Creating the Not-So-Primitive Part

Primitives features are sketch features that use either Extrude or Revolve once you've defined the sketch:

Box   An extruded feature based on a Two-Point Center Rectangle sketching tool. After starting the Box tool, you select a sketch plane or existing face, locate the center, and then set a corner to define the rectangle. You can also set one or two dimensions to precisely size the sketch. Once the sketch is defined, Inventor will automatically begin the Extrude tool, which you use to establish the parameters of the box's size.

Cylinder   The cylinder is another extruded feature using the same basic work-flow as the box. Instead of a rectangle, you use the Center Point Circle sketching tool to define the sketch before launching the Extrude tool.

Sphere   This feature adds functionality to the basic sketch. Initially you use the same Center Point Circle sketching tool you might use with the cylinder, but the Sphere tool will bisect the circle, using half of it and the centerline to define the profile and axis of a Revolve feature.

Torus   This primitive requires the most input to define. The sketch plane that you select will be used to define the section, or the "side," of the feature rather than working from a view of the top. After selecting the sketch plane you will select two points to define the center and the radius of the overall O shape of the torus. Then you will draw a circle using the Center Point Circle sketching tool on the radius endpoint to specify the size of the circle that will be revolved.

Using the Cylinder

Chapter 3 used the box primitive as an example, so in this chapter you will use the other primitives, beginning with the cylinder.

1. On the Quick Access toolbar, click the New icon.

2. Make sure that the 2014 Essentials.ipj is listed as the active project file near the bottom of the New File dialog box.

3. Select the Metric Template set on the left.

4. Locate and double-click the Standard (mm).ipt template file.

5. With the new file open, right-click and select New Sketch from the marking menu.

6. Start the Cylinder tool by using the pull-down in the Primitives panel of the 3D Model tab.

7. Pick the XZ Plane from the `Origin` folder of the browser. Alternatively, you can pick the plane aligned with the x- and z-axes in the Design window.

8. Pick the projected center of the part for the center point of the circle and begin dragging the diameter.

9. Enter 18 for the diameter of the circle and press Enter to begin setting the height of the cylinder.

10. Drag the height of the cylinder to change the value to **30 mm**, as shown in Figure 6.1.

FIGURE 6.1 You can drag the height of the cylinder without having to start the Extrude tool.

11. Right-click and click OK on the marking menu to create the cylinder. This new feature is shown in the Browser as Extrusion1. When using primitives you'll still need to look for Extrusion and Revolution features in the Browser. There is no Box, Cylinder, Sphere, or Torus feature.

▶

Any feature added to the Browser can be renamed to give it more meaning and to make it easier for others to interrogate and edit the model. Using meaningful names is a good practice if you share data with others.

Now change your model view so that you can easily see the front, right, and bottom faces of the ViewCube®.

12. Start the Cylinder tool again by pressing the spacebar or Enter, and select the bottom face of the first cylinder.

By default Inventor will project the edge of the cylinder and the origin of the part into the new sketch once the face is selected

13. Use the projected center of the face to start a new circle with a 10 mm diameter.

14. Set the height of the new cylinder to 20 mm and click OK on the mini-toolbar or marking menu to create the feature.

15. Close the file without saving.

A benefit to using the Cylinder tool in the second case was that it ignored the projected circle and used the profile included in the new Cylinder primitve. If you had simply created a sketch on the bottom of the cylinder you would have had to select which profile or profiles you wanted to create the extrusion from.

Being Well-Rounded with the Sphere

Because the sphere is based on the Revolve tool, you have more flexibility as to what planes you define the sketch on because the feature isn't created in a single direction.

1. Make certain that the 2014 Essentials project file is active, and then open c06-01.ipt from the Parts\Chapter 06 folder.

2. Click the Sphere tool in the Primitives panel of the 3D Model tab.

3. Pick the top of the large cylinder to start defining the sketch.

4. Select the projected center of the face to use as the center point for the circle that will define the sphere.

5. Drag the diameter of the circle to 35 mm, or enter that value in the mini-toolbar. A preview will appear that looks like Figure 6.2.

6. Press the Enter Key to create the sphere.

There are many different ways to create the geometry of this part in Inventor. None of them are wrong. This is part of the reason this type of modeling tool has become so popular.

FIGURE 6.2 A preview of the revolve profile shows it is half of the circle.

Some Uses for the Torus

The Torus is an example of a tool that fits more specific needs. In the following steps, you will use the torus in a couple of ways and be less specific about its precision:

1. Make certain that the 2014 Essentials project file is active, and then open c06-02.ipt from the Parts\Chapter 06 folder.

2. Select the Torus tool in the Primitives panel of the 3D Model tab.

3. Pick the XY Plane from the Origin folder in the Browser.

4. Right-click the plane and select AutoProject from the context menu.

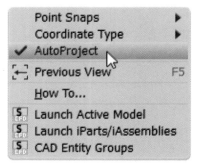

5. If your part does not automatically rotate to align to your sketch, click on the front face of the ViewCube.

6. To use the AutoProject option, move your cursor (or "gesture") over the bottom edge of the large cylinder until a line appears in the sketch.

7. Move your cursor near the middle of the sketch. When you find the middle, it will highlight in green. When this happens, move your cursor up.

   As you move up, an inference line will appear when you are directly over the middle of the line. These inference lines will appear in many situations when you are sketching to offer a reference and, in some cases, build a relationship to existing geometry in the sketch. The coordinates of your cursor will also appear. The value of X should be 0 when the inference line is present.

8. Position your cursor so that X is 0 and Y is roughly 1 mm.

9. Click the mouse to start defining the radius of the torus.

10. Move your cursor to the right until it is equal to the radius of the large cylinder. When the inference line to the edge appears, click to set the radius.

11. The cursor will now switch to setting the radius of the circle that will be revolved. Move your cursor to show that it is tangent with the bottom edge of the large cylinder, similar to Figure 6.3.

12. Click OK to complete the torus.

13. Start the torus again and reselect the xy plane as the sketching plane.

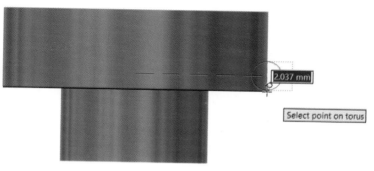

FIGURE 6.3 Inference lines make placing geometry much easier.

14. Hover over the joint where the smaller cylinder connects to the larger to project its diameter into the sketch.

15. Use the same technique to size a torus at the top of the small cylinder, positioning it roughly .5 mm below the large cylinder.

16. Move the radius of the torus to the edge of the small cylinder, set the revolved section radius to come flush with the top of the small cylinder, and click.

17. When the preview appears, change the action to a cut (see Figure 6.4) and click OK to create a relief at the top of the small cylinder.

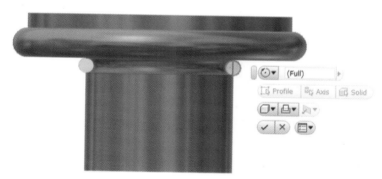

FIGURE 6.4 Using the Torus to create a relief cut

In just a few steps you've created an excellent representation of the part. You can more precisely size the features by simply editing the feature or the sketch that it is based on. You can revisit this part later in the chapter for some additional options, but first let's look at an important subject: work features.

Understanding Work Features

The concept of a work feature is pretty simple: It is a way to create geometry to reference without creating solid or surface geometry to define it. The concept stands in contrast to the fantastically broad array of applications work features have.

Work features can be created using all of the fundamentals of planes, axes, and points. For example, a plane can be placed between axes, three points, or normal (perpendicular) to an axis at a point intersecting it. An axis can be created to pass through two points or at the intersection of two planes.

Inventor builds on these geometry fundamentals. For example, you can add a work plane using the midpoint of an edge and the edge itself, or by creating "inline" features to aid in creating a work feature. This is where you can start the Work Plane tool and then create the two axes to define it.

In the next several exercises, you will create planes and axes in many ways. A key to understanding these tools for your own work is to explore and consider *geometric dissection*. This is a process of breaking complex shapes down to their core to see how they were constructed. You will not use every placement option listed in the pull-down menus (Figure 6.5), but we will cover some of the most commonly used options. In most cases, through careful geometry selection you can get the same result as a pull-down option by using the basic tool.

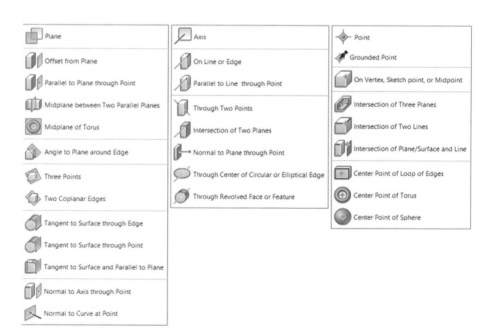

FIGURE 6.5 The options for placing a Work Plane, Axis, or Point

The Midplane Work Plane

So far nearly every part that you have created has been centered on the x-, y-, and z-axes of the part file. If you do not construct your parts this way and need to pass a plane through the part's center, the Midplane Work Plane is the tool for you. It can also be used to center a plane between any two parallel faces.

1. Make certain that the 2014 Essentials project file is active, and then open c06-03.ipt from the Parts\Chapter 06 folder.

2. Click the Plane tool in the Work Features panel of the 3D Model tab or click the] key.

3. Pick the Blue face in the partial slot in the part.

4. You can rotate your part using the Orbit tool or ViewCube to pick the red face, or you can simply hover your cursor where you think the opposite face in the slot is to highlight and pick it. See Figure 6.6.

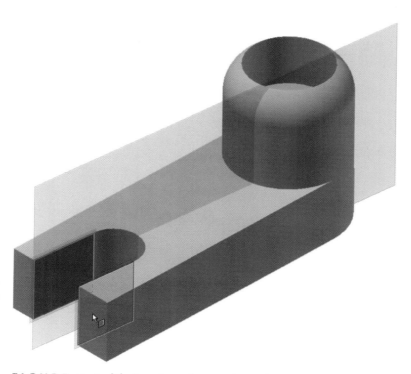

FIGURE 6.6 Select one face, edge, or point, and work features will automatically eliminate others from being selected.

5. The tool will automatically stop once it has created the plane. Rotate the part to see that the plane bisects the slot. Note that the sides of the slot are not parallel to the rest of the part.

   Work features are associative to the geometry that created them. A change to their definition will update them.

6. Find the extrusion named Slot in the Browser or pick on the red or blue face and edit the sketch.

7. Change the angle in the sketch from 85 to **70**.

8. Finish the sketch and see that the angle changes with the part update.

   A pull-down menu option is available for creating this type of plane that will automatically limit options other than selecting parallel planes.

9. Move your cursor near the edge of the new work plane; when it highlights, right-click and choose Visibility from the marking menu to deselect its visibility so it is hidden in the Design window.

Turning off a work feature's visibility does not affect it or any geometry based on it. This is a good technique to use to remove clutter from the Design window as you build complex models.

Normal to a Curve at Point

You can create a point along a curve in a several ways in Inventor, and you will do so later in this chapter. But creating a plane normal to a curve at its endpoints or midpoint does not require you to make an actual work point. In these steps, you will create a plane that is perfect for locating a sketch to be used for a Sweep profile.

1. Make certain that the 2014 Essentials project file is active, and then open c06-04.ipt from the Parts\Chapter 06 folder.

2. Use the pull-down under the Plane button in the Work Features panel to select the Normal to Curve at Point option.

   Note that as you move over faces on the part they do not highlight. Only natural points, work points (there are none in this part), and curved edges highlight.

3. Move your cursor along the upper long edge to highlight its midpoint, as shown in Figure 6.7, and pick the point.

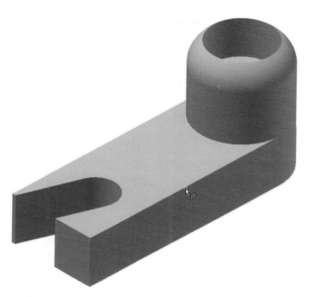

FIGURE 6.7 Using naturally occurring points is quicker than creating points explicitly.

4. The edge that passes through the point should highlight and preview the plane that can be created. Select the edge to create the work plane.

This work plane construction is very common and a similar tool (Normal To Curve At Point) works on complex curves. Now you will create a series of planes and a work axis to define the one you need for placing a sketch.

Stacking Features

There are times when one or two faces or a couple of points will not give you the geometry that you need. In these cases, you can build multiple work features to lead up to the one you really need:

1. Make certain that the 2014 Essentials project file is active, and then open c06-05.ipt from the Parts\Chapter 06 folder.

2. Start the Axis tool in the Work Features panel or press the / key.

3. Pick the large cylindrical face to place that axis at the center of it.

4. Select the Angle To Plane Around Edge tool from the Work Plane pull-down.

5. Pick the work axis you created; then pick the front face of the part.

The plane is infinite, so changing its display doesn't affect its functionality. By default the size of the plane display is based on the size of the body it intersects. You can also control this by enabling or disabling the Auto-Resize option on the context menu. ▶

6. When the preview of the plane appears, drag the angle arrow or set the value through the mini-toolbar to 30. See Figure 6.8.

7. Click OK to create the new feature.

   You can use an existing edge to create the same effect, but round features are common sources for a datum on a part so attaching a work axis to one of those features and building from it often makes great sense.

8. Move your cursor near one of the corners of the new work plane until a double-headed arrow glyph appears (a quadruple arrow will move the plane).

9. Click and drag the corner (and others) to resize the displayed plane.

   If you wanted to create a parallel plane a certain distance from this plane, you could use the Offset Plane option. In this case you will create a plane that is parallel but its offset distance is based on the radius of the cylinder you placed the axis through. To be able to follow this distance, you will create a new plane that is tangent to the old one.

10. Select the Tangent To Surface And Parallel To Plane option from the Work Planes pull-down.

11. Select the cylindrical surface and then the last plane you created at an angle to create the new work plane.

12. Use the Offset From Plane option to create a plane that is offset 15 mm from the last tangent plane, as shown in Figure 6.9.

13. Click OK to create the plane.

    Now you have a number of planes visible on the screen that you may not want to see all of the time. You can change their visibility individually, or you can use a keystroke switch to turn them off.

14. Hold the Alt key and press the] and / keys to turn the user-created planes and axes on and off.

    Ctrl+] and Ctrl+/ will toggle system-created plane and axis visibility. Ctrl+. (period) will do the same for work points.

Now let's look at options for incorporating sketches into creating work features.

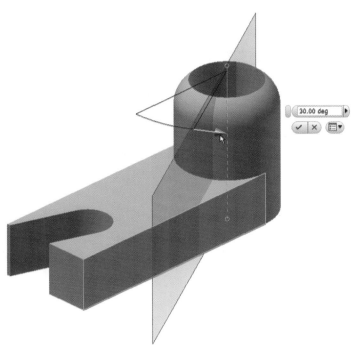

FIGURE 6.8 Creating a plane through an axis at a specific angle

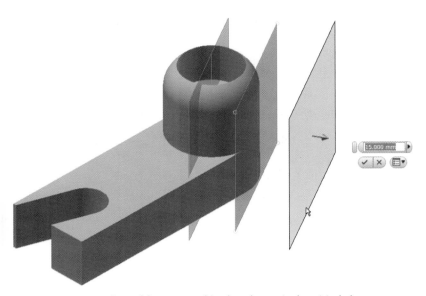

FIGURE 6.9 Several features combined to place a single, critical plane

Sketch-Based Work Features

Sketches are powerful tools for creating 3D features. They can be used by work features in the same way, and the work features will respond to changes in the sketch as they do with changes to 3D geometry.

1. Make certain that the 2014 Essentials project file is active, and then open c06-06.ipt from the Parts\Chapter 06 folder.

2. Select the Angle To Plane Around Edge option from the Work Plane pull-down or simply start the Work Plane tool.

3. Pick the 25 mm line at the 80 angle in the sketch and then pick the front face of the part (see Figure 6.10).

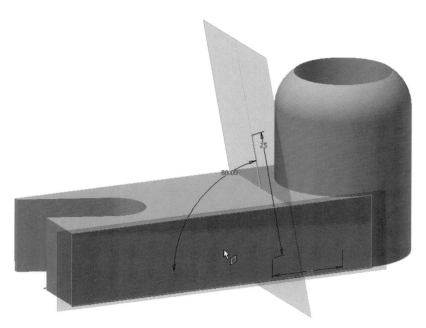

FIGURE 6.10 Creating a plane that has its angle controlled by a sketch

4. Leave the angle at 90 and click OK to create the work plane.

5. Use the browser to turn off the visibility of Sketch 4 and the new work plane.

6. Right-click the End Of Part marker and click Move EOP To End.

▢	<u>R</u>epeat Work Plane
🔲	Move EOP to <u>T</u>op
🔲	Move EOP to <u>E</u>nd
	<u>D</u>elete All Features Below EOP
	How To...
[S]	Launch Active Model
[S]	Launch iParts/iAssemblies
[S]	CAD Entity Groups

In the previous exercise the idea was that you could offset a plane from the angled plane if you knew the complete offset distance. In the following example, you will use a sketch to set the distance and angle of the new work plane without creating interim features.

7. Right-click and select the Work Plane tool from the marking menu.

8. Pick the endpoint of the 25 mm sketched line when it highlights.

9. Pick the line (Figure 6.11) to use as the axis the plane will be normal to.

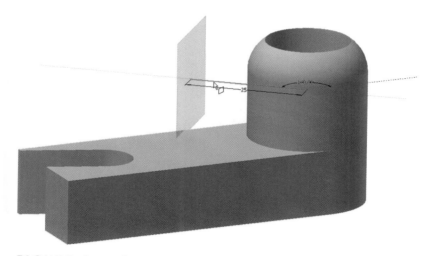

FIGURE 6.11 Creating a plane normal to a path

This creates the plane based solely on the sketch. This sketch is based on other planes of faces, which create their own dependencies — all of which can be controlled and, if need be, redefined.

Combining Various Work Features

In Chapter 3 you created a hole based on a work point and a work axis. Now you will use various work features to begin construction of a similar point and axis using planes, axes, and a surface. This will take a few steps, so please read carefully:

1. Make certain that the 2014 Essentials project file is active, and then open c06-07.ipt from the Parts\Chapter 06 folder.

2. Select the Plane tool in the Work Features panel of the 3D Model tab.

3. Select the y-axis of the part from the Browser or the Design window; then select the Blue face of the slot. Leave the angle of the face at 90 and click OK to create the new work plane.

4. Right-click and choose Work Plane from the marking menu to create another plane.

   Next you will create an axis that is aligned with this new plane at a distance above the yellow face on the part. To do so, you can create an offset plane from the yellow face ahead of time, or you can create a plane "in line," defining it while in the Work Axis tool. This is the technique you will use.

5. Start the Axis tool from the 3D Model tab's Work Feature panel.

6. Right-click and select Create Plane from the context menu.

   This will temporarily start the Work Plane tool. The plane you define will be used to intersect another plane to create an axis.

7. Click and hold the yellow face on the part. Drag a plane 8 mm above the yellow face of the part (see Figure 6.12).

8. Click OK to create the in-line plane; then select the first work plane you created in this exercise to generate the work axis.

   The plane you just created will be embedded under the work axis. It can be used for other features, but because it was created as part of the work axis, it will automatically be hidden.

9. Start the Work Plane tool again; select the work axis and then the yellow face.

10. Set the angle of this new work plane at 45 degrees and click OK to create the plane. Compare your results with Figure 6.13.

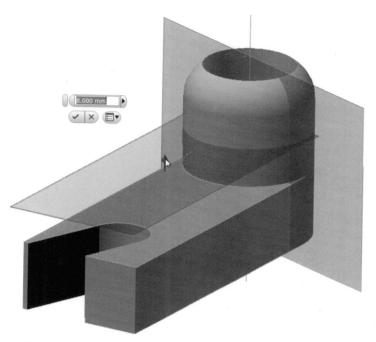

FIGURE 6.12   Creating a new offset work plane as part of defining a work axis

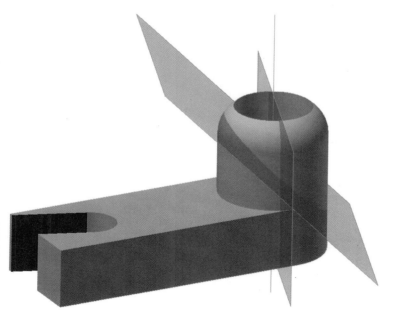

FIGURE 6.13   Work features can be used to build other work features.

11. Start the Work Axis tool and once again right-click and select the Create Plane tool from the context menu

12. Select the y-axis again and then click the first work plane you created in this exercise.

13. Set the angle to 90 and click OK to create the plane.

At this point you have created most of the planes necessary to locate a work point and have it be controllable. In the next exercise you will complete the construction of the work point and work axis you need.

Completing the Construction

In the previous exercise, you set up the position and control of the two work planes needed to define the work axis. In this exercise you will use the work features you already created to construct a Work Point.

1. Make certain that the 2014 Essentials project file is active, and then open c06-08.ipt from the Parts\Chapter 06 folder.

2. Start the Work Axis tool from the Work Features panel of the 3D Model tab or press the / key.

3. Pick the two work planes that are visible in the Design window, and the work axis will be created at their intersection.

4. Press Alt+] key to turn the work planes off.

5. Press the . (period) key or click the Work Point tool in the Work Features panel of the 3D Model tool.

6. Pick the new work axis and the curved face shown in Figure 6.14 to create the work point.

In the next section, you will focus on creating features using additional options. Then you will see how editing work features after they've been used to create geometry can affect the part.

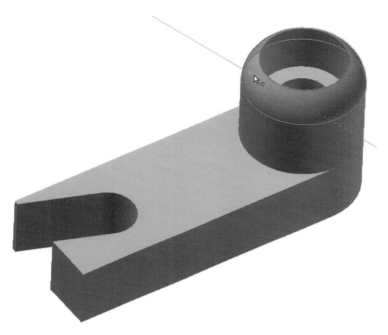

FIGURE 6.14   A work point will be created at the intersection of the axis and the curved face.

Using Sketched Feature Termination Options

The creators of Inventor have done a very good job offering consistency between features in their interface. In this section, you will get to see what some of these options look like in practice.

You will not be using every option for every feature. The use of an option for the Extrude tool will be very similar to the same option put to use in the Revolve tool.

These options are important, though, and understanding how they work can affect how you go about setting up your model to be able to use an option to save you steps for creating your designs.

Using To and To Next

In previous chapters, the extrusions you have used were done to a specified distance. In this exercise, you will compare the function of two common substitutes to extruding to a distance.

1. Make certain that the 2014 Essentials project file is active, and then open c06-09.ipt from the Parts\Chapter 06 folder.

2. Right-click in the Design window and pick the Extrude tool on the marking menu.

3. When Extrude starts, it will pick the available profile and begin to create a feature to a distance.

4. Use the Termination Options pull-down to switch from Distance to To Selected Face/Point.

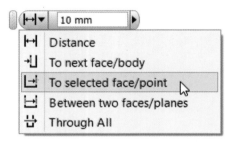

5. Select the cylindrical face on the outside of the part.

6. Click OK to create the feature.

   Review the feature as it was created. If you look closely at the part, you will see that the extrusion left a gap at the top and, when it terminated, it even followed the contour of the surface. See Figure 6.15.

7. Pick a face of the new feature. Pick the Edit Extrude tool from the mini-toolbar.

8. When the feature editing tools appear, change the termination to To Next Face/Body.

   The preview of the feature will update to show the extrusion now wraps onto the face of the fillet.

9. Click OK to update the feature to look like Figure 6.16.

Both termination options in this exercise are used for a number of different things. Another popular approach is to embed a sketch within the part and create a feature out to a work plane.

FIGURE 6.15 Terminating to a face gives the feature the contour of the selection.

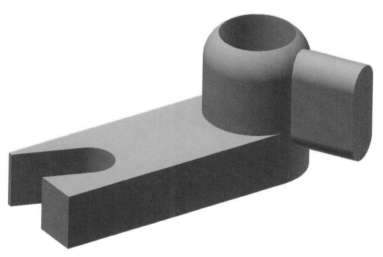

FIGURE 6.16 When features terminate on the next available face, they match that geometry.

Building between Faces

Rather than constructing a sketch face and a To face to create a feature, if you have faces that a feature would fit between, you can have a sketch just about anywhere to create a feature.

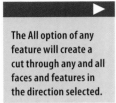

The All option of any feature will create a cut through any and all faces and features in the direction selected.

1. Make certain that the 2014 Essentials project file is active, and then open c06-10.ipt from the Parts\Chapter 06 folder.

2. Start the Extrude tool and switch the termination option to Between Two Faces/Planes.

3. Select the two visible work planes (Figure 6.17) and click OK to generate the feature.

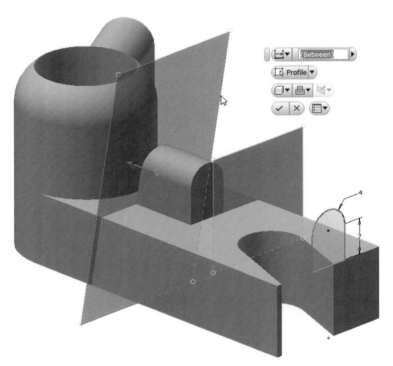

FIGURE 6.17   Sketches can be held away from the termination faces or between them.

The feature is created normal to the plane that the sketch was on. What angles of the planes or curvatures of the termination faces that exist will be incorporated into the feature, but that affects only the ends of the feature.

Building Features in More Than One Direction

To this point you have been working with termination options for features created in a single direction. Inventor has useful options for creating geometry in more than one direction.

1. Make certain that the 2014 Essentials project file is active, and then open c06-11.ipt from the Parts\Chapter 06 folder.

2. Right-click in the Design window and pick Extrude from the marking menu.

3. Change the direction on the mini-toolbar to Symmetric.

4. Drag the arrow to set the height of the extrusion to **30 mm**.

5. Click and drag the ball at the base of the arrow and see how draft is added to the faces above and below the sketch plane.

6. Set the draft angle to –2 degrees (see Figure 6.18).

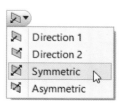

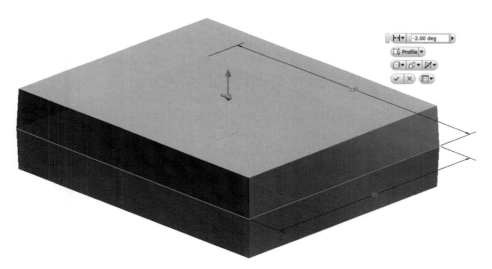

FIGURE 6.18 A symmetrical termination divides a feature evenly on either side of the sketch plane.

7. Press Enter to create the feature.
   The block is created with draft on all of the faces. The height of the extrusion is the same on either side of the sketch plane.

8. Drag the End Of Part icon to the bottom of the Browser.
   This will expose another sketch.

9. Start the Extrude tool and switch the termination to Asymmetric.

10. Set the first Distance value to **25 mm** and the second to **10 mm**.

11. Click the More tab and set the Taper angle value for both distances to –5 degrees to create the feature shown in Figure 6.19.

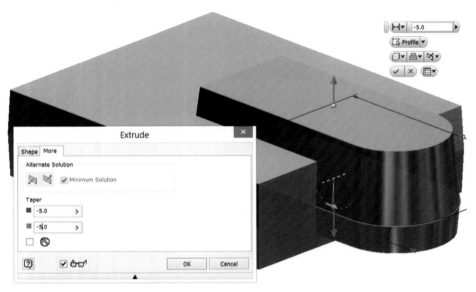

FIGURE 6.19 Asymmetric termination offers flexibility on either side of the sketch plane.

12. Click OK to create the new feature.

Asymmetric termination allows for unique distance and taper angle for each direction in the Extrude tool and allows for unique angles for revolved features.

Editing and Redefining Work Features

Work features offer an efficient way to create features that would be difficult to define otherwise. The creation of work features is relatively simple, but like 3D features, their real power lies in their ability to be edited without the destruction of the components.

Modifying a Plane to Change a Feature

Updating the plane that a sketch is dependent on will cause the feature to be recalculated, as seen in the following exercise:

1. Make certain that the 2014 Essentials project file is active, and then open c06-12.ipt from the Parts\Chapter 06 folder.

2. Locate the feature labeled Offset Plane in the Browser and double-click its icon.

3. The features dependent on this work plane will disappear and the value used to create the plane will be visible on your screen.

4. Change the Offset value from **15 mm** to **10 mm** and click OK to update the model.

5. Expand the features that make up the offset plane in the Browser: the plane labeled Tangent Plane, Plane At Angle, and the part's y-axis.

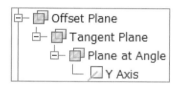

6. Double-click the Plane At Angle feature to show its construction.

7. Change the angle of the plane to **60** degrees. The updated model is shown in Figure 6.20.

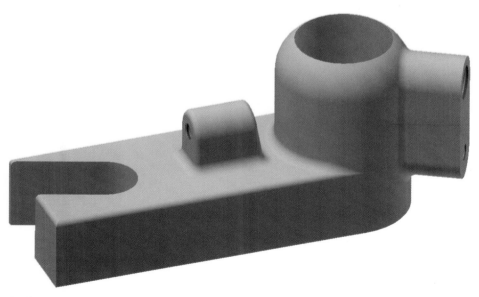

FIGURE 6.20 The updated boss shortened and rotated by editing work planes

Depending on the severity of the change to your model, you may need to update features dependent on the edited work features. Editing features simply changes the properties of their definition. There are times when you need to rethink how you created the feature in the first place. To do so, you must redefine the feature.

8. In the Browser, right-click the work plane labeled Normal To Edge At Point and select Redefine Feature from the context menu.

   This feature was originally created at the midpoint of the model edge normal to that edge. Now you will make it a plane set a distance from the end of the part.

9. Pick the rectangular face at the end of the part; then click and drag the plane 22 mm from the end of the part, as shown in Figure 6.21.

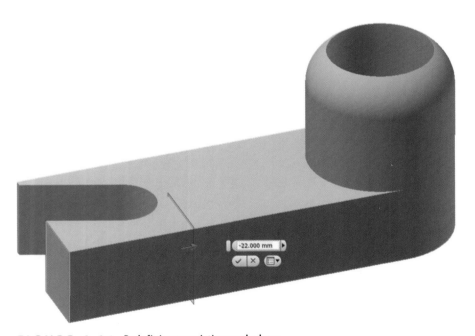

FIGURE 6.21 Redefining an existing work plane

This will cause the feature created between two work planes to update its length. The tapped hole will also change its length.

The use of work features is universal regardless of the type of part you are creating. Most of the termination options appear in every kind of sketched feature whether they are extrusion, revolved feature, or loft features.

THE ESSENTIALS AND BEYOND

Primitives offer a different way of developing your parts. Learning to "dissect" complex components into primitives will open you to new approaches with your models.

Using work features to develop the foundation for primitives and other sketched features is a critical skill to develop to create complex models.

ADDITIONAL EXERCISES

▶ Look at 3D parts around you and try re-creating them using primitive features.

▶ Work on combining work features in-line rather than explicitly to define other features.

▶ Practice selecting work features and editing them through the in-place menus like you do with sketched and placed features.

CHAPTER 7

Advanced Part Modeling

Now that you've developed the basic skills for part modeling and explored some additional foundational knowledge, it is a good time to let you know that you've already learned some of the most challenging concepts in Autodesk® Inventor® 2014. Advanced features are as easy to define and often easier to construct than the basic features.

In this chapter, you will create or complete a handful of parts that include complex contours or large numbers of features. As you move beyond basic, prismatic parts, you need to work with more developed sketched and placed features. You will find that the basics learned with primitives, extrusions, and revolved features will apply directly to the dialog boxes and workflows of these more advanced tools.

In addition, by completing this chapter and Chapters 3 and 6, you will have used the majority of solid modeling tools available in Inventor.

▶ **Creating 3D sketches**

▶ **Building a hole pattern**

▶ **Exploring advanced efficiency features**

Creating 3D Sketches

Certification
Objective

For your first challenge, you will construct a spoke for a hand wheel that will change shape as it passes along a path. To do this, you will use a loft feature, but first you need to develop some of the elements required to form the feature.

Projecting a 3D Sketch

You can use a 3D sketch to create complex geometry for sketched features. It can be created with a stand-alone tool or as an option to a 2D sketch that defines part of it.

To define the second profile, you need to build a plane on a curved face, which you will do in the following steps:

1. Make certain that the 2014 Essentials project file is active, and then open the c07-01.ipt file from the Parts\Chapter 07 folder at the book's web page, www.sybex.com/go/inventor2014essentials.

2. Double-click Loft Sketch 2 in the Browser to edit the sketch.

3. Locate the Project Geometry tool in the Draw panel of the Sketch tab. Expand it, and start the Project To 3D Sketch tool.

4. In the dialog box, select the Project check box (Figure 7.1), click the curved face, and click OK.

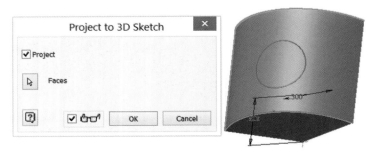

F I G U R E 7 . 1   Projecting a 2D sketch onto a 3D face

   This will create a new 3D sketch and apply the circle to the cylindrical face.

5. Click Finish Sketch in the Ribbon or use the marking menu.

Creating the 3D sketch has given you a second profile for your loft. Now you need a path for the two profiles to transition along.

Defining a Loft Path Using a 3D Sketch

A 3D sketch can be constructed as a series of lines and an arc directly in the Design window. A technique that can be even easier to master is to use the

Intersection Curve tool to create a new sketch by combining two 2D sketches, two surfaces, or a combination of the two.

In this exercise you will create a 3D sketch for your loft path by combining two existing sketches:

1. Make certain that the 2014 Essentials project file is active, and then open c07-02.ipt from the Parts\Chapter 07 folder.

2. From the drop-down menu in the Sketch panel on the 3D Model tab, select Create 3D Sketch.

Create 3D Sketch

3. Click the Intersection Curve tool in the Draw panel of the 3D Sketch tab.

4. Click the Red and the Blue sketches in the Design window and click OK.

Intersection Curve

5. Finish the 3D sketch using the Finish Sketch tool.

6. Turn off the visibility of Rail Sketch 1 and Rail Sketch 2 in the Browser. Your model should look like Figure 7.2.

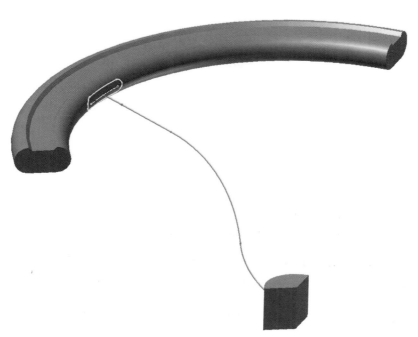

FIGURE 7.2   The sketch of the loft path

The sketch consists the combined lines and arcs of the 2D sketches. Now that you have the pieces in place, you can create your loft features.

Creating Loft Features

Loft features are created using the Loft tool, which offers several options. For example, you can define whether sketches or model faces are used for profiles (sections) and whether the paths will connect to the geometry. You will now use this tool to create a wheel spoke:

1. Make certain that the 2014 Essentials project file is active, and then open c07-03.ipt from the Parts\Chapter 07 folder.

2. Start the Loft tool from the Create panel of the 3D Model tab.
   The dialog box that opens (Figure 7.3) displays two primary areas for sections and rails or centerlines. As you click geometry, the selected parts will appear in these windows.

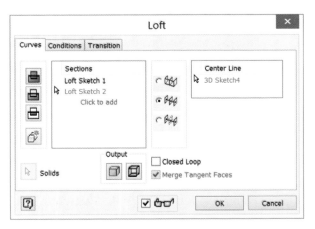

FIGURE 7.3 The Loft dialog box showing the Center Line path option

When you start the tool, it will select sections first.

3. For the first profile, click the yellow sketch on the rim of the wheel segment. It is labeled as Loft Sketch 1 in the Browser, or you can select it from the Design window.

4. Click Loft Sketch 2 for the second profile.

5. Between the Sections and Rails groups are the Rail options. Set the option for Center Line.

6. On the Curves tab, click where it says *Select a sketch* in the Center Line group. Select the 3D Sketch path.

   As you've seen with other sketched features in Inventor, a preview will appear when all of the elements needed to define it are satisfied.

7. Once a preview looks like Figure 7.4, click OK to create the feature.

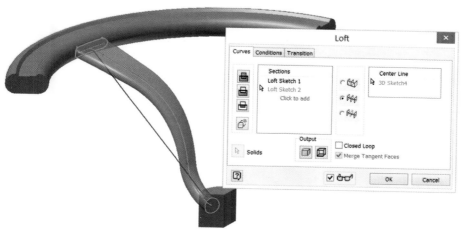

FIGURE 7.4 A preview of the loft feature will appear after the sketches are selected.

8. Select the Circular tool from the Pattern panel of the 3D Model tab.

9. In the Circular Pattern dialog box, pick the Pattern A Solid option.

10. Pick the edge in the middle of the hub for the rotation axis and set the number of placements to 3.

11. Click OK to generate the feature (see Figure 7.5).

The loft has created a connection between the hub and rim of the wheel. This examples used the minimum of two profiles, but you can add as many profiles as you need to accurately develop the shape you want. One loft option makes it possible to control the area of each section for more advanced designs.

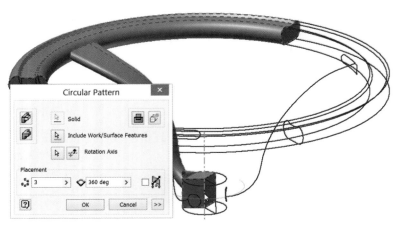

FIGURE 7.5   Patterning the entire solid allows you to edit a smaller feature set.

Using the Engineer's Notebook

To assist others who might need to edit your models in the future, it can be helpful to document the process or thinking behind a design. Inventor has a tool called the Engineer's Notebook that enables you to capture comments or images of your design at any time and keep them with the component for future reference.

1. Make certain that the 2014 Essentials project file is active, and then open c07-04.ipt from the Parts\Chapter 07 folder.

2. Double-click the Note icon in the Design window to open the notebook (see Figure 7.6) included in this file.

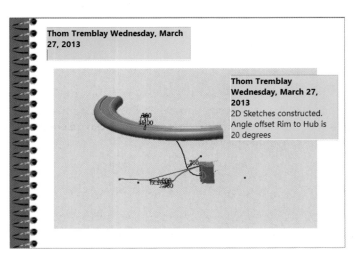

FIGURE 7.6   The contents of the notebook in the file show its history.

3. Right-click in the open notebook and choose Comment from the marking menu, or select the Comment tool from the Notes panel of the Engineer's Notebook tab.

4. Draw the rectangle that forms the comment, and add text similar to Figure 7.7.

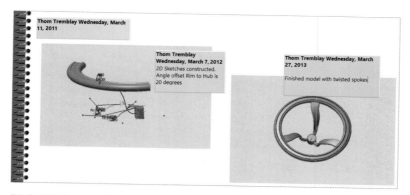

FIGURE 7.7 Be sure to include a good description of changes in text and images.

5. Pick View from the marking menu or from the Notes panel, and create a view of the Graphics window.

   By right-clicking and using the context menu, you can rotate, pan, or zoom the view to get a better image, as shown in Figure 7.7.

6. Right-click on the image and select Freeze from the context menu to preserve the view at this point in the model development for future reference.

7. Select Finish Notebook from the Exit panel to return to editing the model.

T I P It is possible to create more than one Engineer's Notebook in a single component for different reasons. Notes can be created at any time and can be renamed or removed.

Creating a Sweep with a Guide Rail

To create a sweep, you define a profile, a path, and, optionally, a *guide rail* or *guide surface*. For this exercise, you will use the optional guide rail, which will

Certification
Objective

steer the profile as it sweeps along a path. Follow these steps to use the guide rail option:

1. Make certain that the 2014 Essentials project file is active, and then open c07-05.ipt from the Parts\Chapter 07 folder.

2. Locate the Guide Path sketch in the Browser; right-click and select Edit 3D Sketch from the context menu.

3. Expand the 3D Sketch in the Browser and double-click on the Helical Curve1 feature to open the Helical Curve dialog box.

4. Change the Revolutions value to .25 and click OK to update the curve; then click Finish 3D Sketch.

   With the 3D sketch updated, you can use this sketch to create a specialized sweep.

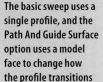

5. Start the Sweep tool from the Create panel of the 3D Model tab.

   Since there is only one closed profile in the model, it will automatically be selected for the profile.

6. Select the 120 mm line as the path. This preview will look like an extrusion.

7. Change Type from Path to Path & Guide Rail.

   When you change the option, an icon appears with a red arrow, which indicates that Inventor needs input from the user. You must satisfy all of these red arrows to complete the feature.

8. Select the curved line that is part of a 3D sketch as the guide rail.

9. Click OK to create the new feature, as shown in Figure 7.8.

The use of the 3D sketch's helical curve allows the profile to twist around the path. A surface can be used to add control in different situations.

Creating a Sweep with a Guide Surface

The previous exercise showed how using a guide rail can make it easy to create something complex. Sweeping a polygonal profile along a curved rail can sometimes cause the profile to twist uncontrollably.

In this exercise you will see how using a guide surface can keep a profile in its proper attitude to the path:

> Helical curves can be defined using combinations of revolutions, height, or pitch. The feature can also create flat spirals.

> The basic sweep uses a single profile, and the Path And Guide Surface option uses a model face to change how the profile transitions along the path.

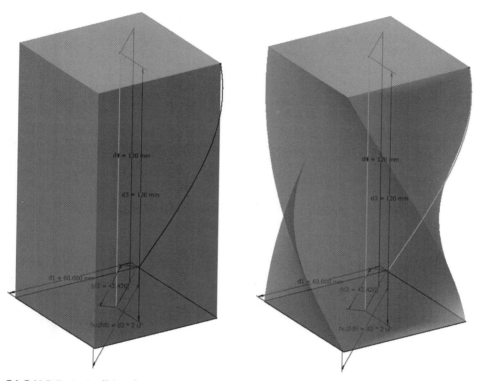

FIGURE 7.8 Using the guide rail allows the feature to be more complex.

1. Make certain that the 2014 Essentials project file is active, and then open c07-06.ipt from the Parts\Chapter 07 folder.

2. Start the Create 3D Sketch tool in the Sketch panel of the 3D Model tab.

3. Select the Project To Surface tool from the Draw panel on the 3D Sketch tab.

   The Project To Surface tool offers several options for controlling how 2D sketch data is projected onto a 3D face. The sketch you will use was drawn as though the path's centerline was drawn to its desired total length in the sketch plane.

Project to Surface

4. Click the Wrap To Surface icon in the Project Curve To Surface dialog box.

5. Pick the cylindrical face of the part; then pick the Curves Selection icon.

6. Drag a window around the sketch geometry in the Design window to select the two lines and one arc that are in the sketch.

7. Click OK to create the 3D sketch and then finish the sketch.

8. Turn off the visibility of Sketch2 in the Browser.

9. Move the End Of Part marker in the Browser below the Profile sketch that is suppressed. Your model should look like Figure 7.9.

FIGURE 7.9 The projected path and profile ready for use

10. Start the Sweep tool from the Create panel on the 3D Model tab.

11. The profile is automatically selected, so pick your newly created 3D sketch for the path.

    Take a moment to look at the preview. If you turn the model so you can see the left and bottom faces on the ViewCube®, you will see that the profile of the sweep is twisting.

12. Change Type to Path and Guide Surface.

13. Select the cylindrical surface of the part and see that the profile now transitions with the curvature.

14. Switch the feature to Cut Material; then click OK to create the groove shown in Figure 7.10.

FIGURE 7.10   A guide surface keeps the profile properly aligned.

The depth and rotation of the feature will now be consistent with the profile. See Chapter 3, "Introducing Part Modeling," for a more common use for the Sweep tool: sweeping a profile along a path.

Creating a Shell

When parts are hollow with a consistent wall thickness along a face, using the Shell tool is much more efficient than building the walls separately.

In the following steps, you will use the Shell feature and its options to modify more than one body in a single part:

Certification
Objective

1. Make certain that the 2014 Essentials project file is active, and then open c07-07.ipt from the Parts\Chapter 07 folder.

2. Start the Shell tool from the Modify panel of the 3D Model tab.

 Shell

The Shell dialog box (Figure 7.11) can be expanded for additional options. You will be prompted to click the faces you want removed to create the hollowed shape and to name a thickness for the remaining walls.

FIGURE 7.11 Expanding the Shell dialog box gives you access to additional wall thicknesses and options.

3. You want to remove the green face, so first select it.

   Doing so will preview how the feature will hollow out the part, removing the green face and any face tangent to it. In this case, you only want to remove the green face.

4. Press Esc to cancel the Shell tool and then press Enter to restart it.

5. Deselect the Automatic Face Chain option in the Shell dialog box, and then select the green face on the part again.

6. Set the Thickness value to 5 mm by dragging the arrow or entering the value in the dialog box.

7. Expand the dialog box by clicking the More icon on the bottom of the dialog box.

8. Click where the dialog box says *Click To Add*; then click the yellow face on the part, and set the value to **15 mm**.

9. Click OK to create the shell feature, as shown in Figure 7.12.

After starting the Shell tool you might notice a preview on your part showing it hollowed out but having no faces removed. This is an option available to you. You can use a feature to open the part or with additive manufacturing, an enclosed void is now acceptable.

F I G U R E 7 . 1 2 The resulting shell shows the difference in wall thickness.

10. In the Solid Bodies folder in the Browser, turn off the visibility of the Block body and turn on the visibility of the Pin body.

11. Start the Shell tool again and turn the Automatic Face Chain option back on using the mini-toolbar icon.

12. Select the bottom face of the pin to remove this face.

13. After the preview is generated, change the Shell solution to Outside.

14. Drag or enter the value of the shell to 13 and click OK to create the feature.

The wall thickness of the part exceeded the minimum size of a fillet on the part, so Inventor removed the fillet and left a sharp edge in its place. This is an important feature of the Shell feature. It is also something to be aware of when using small fillets because you might need to add fillet features to the part to avoid issues created by sharp edges.

15. Pick the sharp edge at the meeting point of the two cylinders and add a 10 mm fillet, as shown in Figure 7.13.

FIGURE 7.13 Shell features can also work in reverse.

You can also create a shell feature that adds material on either side of the original model face.

Building a Hole Pattern

In Chapter 3, you replicated a hole feature around an axis using a circular pattern. Holes can also be created in patterns that space them along a linear direction.

Creating a Rectangular Hole Pattern

Let's create a series of holes. On this part, the holes will be in a pattern where the two directions are perpendicular, but the pattern can go in any two directions.

Certification
Objective

1. Make certain that the 2014 Essentials project file is active, and then open c07-08.ipt from the Parts\Chapter 07 folder.

2. In the Pattern panel on the 3D Model tab, click the Rectangular Pattern tool.

3. The dialog box requires you to select a feature, so click Hole6 in the Browser.

4. Click the button with the red arrow in the Direction 1 group, and then click the short edge shown in Figure 7.14.

5. Set the number of instances to 3 and the distance between them to 34.

6. Click the Direction 2 icon, and select the longer edge shown in Figure 7.14.

7. To set the proper direction of the new instances, click the Flip button.

8. Set the number of instances to 2 and the distance to 52 as shown in Figure 7.14.

FIGURE 7.14 Creating a pattern of holes

9. Click OK to place the pattern on the part.

   One of the occurrences in the pattern is not needed; it interferes with other features of the part.

10. Locate the Rectangular Pattern1 feature in the Browser, and expand it.

11. Hover over the Occurrence icons until the hole that overlaps the notch in the part highlights, as shown in Figure 7.15.

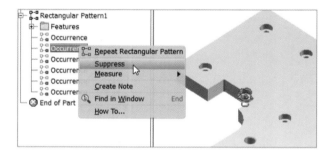

FIGURE 7.15 Individual instances in a pattern can be suppressed.

12. Right-click and click Suppress on the context menu. The completed feature will look like Figure 7.16.

FIGURE 7.16 The hole pattern with one occurrence suppressed

It is possible to suppress any number of members and to change whether they're suppressed at any time.

Expanding on the Rectangular Pattern

The fundamentals of the Rectangular or even the Circular pattern are much the same. As mentioned in the previous exercise, the directions of the pattern don't have to form a rectangle.

In this exercise, not only will you use one direction, but you will see that direction can be a curve.

1. Make certain that the 2014 Essentials project file is active, and then open c07-09.ipt from the Parts\Chapter 07 folder.

2. Start the Rectangular Pattern tool from the Pattern panel of the 3D Model tab.

3. Select the Hole Boss and Hole1 features from the Browser.

4. In the Rectangular Pattern dialog box, click the Direction 1 selection icon and pick the projected sketch that runs along the edge of the opening.

5. Set the number of instances to 6, and use the drop-down to change the spacing to Curve Length, as shown in Figure 7.17.

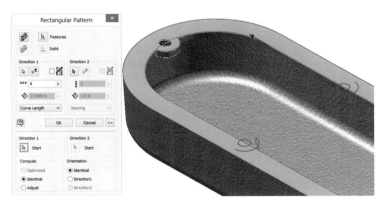

FIGURE 7.17 Be sure to select the sketch for your path and the hole center for a start point.

The initial preview shows instances of the features in space. It is usually necessary to define the start point of the pattern to make sure it follows the curve properly.

6. Expand the dialog box, click the Start Of Direction 1 icon, and then select the point at the center of the hole.

7. When the preview updates showing the instances following the sketch, click OK to create the new features.

   The new hole bosses and holes will be created, but there is one more change to make to the model before you continue.

8. Expand the Extrusion1 feature in the Browser, and edit Sketch 1.

9. Change the 2.5-inch dimension to 4 inches, and finish the sketch. To see what the updated model will look like, take a look at Figure 7.18.

FIGURE 7.18 Updating the part updates the sketch and therefore the pattern.

Using a sketch as the path for a pattern opens a lot of options for creating complex patterns easily.

Exploring Advanced Efficiency Features

Among the advanced features are editing tools that enable you to increase efficiency by replicating features, editing features after their creation, and even replacing a portion of an existing feature to give it a new contour.

In the next group of exercises, you will convert a simple block into a complex bottle using simple tools. You will see how a good workflow allows the complex problem to be broken down into a few simple steps.

Combining Fillet Types

Placing individual types of fillets is an easy process, but sometimes combining types of fillets can allow you to build complex blends more easily and accurately than by applying them one step at a time. That's how you'll begin to add some nice shape to your bottle.

1. Make certain that the 2014 Essentials project file is active, and then open c07-10.ipt from the Parts\Chapter 07 folder.

2. Orbit the part so that you can see the right and top faces.
   For this exercise, you will use the names of the ViewCube faces to identify the faces on the part and the edges where they intersect.

3. Start the Fillet tool, which is in the Modify panel of the 3D Model tab or in the marking menu.
   When the tool starts, a dialog box and a mini-toolbar will appear. For the exercise, you will focus on the mini-toolbar.

4. In the mini-toolbar, use the drop-down to change the fillet type to Full Round Fillet.

5. Click the right, top, and left faces of the part. As soon as you click the third, the preview of the fillet will appear.

6. Move back toward the mini-toolbar, and it will expand. Click the Apply icon (the green plus sign) to add the fillet, and continue selecting other options.

7. On the mini-toolbar, change the fillet type to Edge Fillet.

8. Return to the Home view, click the edges shown in Figure 7.19, and set the radius to 3. Do not click OK or Apply.

9. In the mini-toolbar, change the type of edge fillet to Add Variable Set, and click the edge shown in Figure 7.20.

10. Click near the middle of the curved edge to place an additional radius reference point.

FIGURE 7.19   Place a constant radius on the model edges.

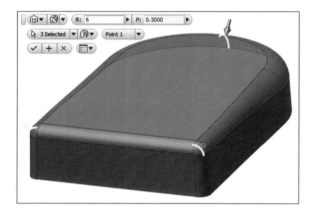

FIGURE 7.20   Add a point along the edge to add another radius.

11. Set the radius to 6 and the edge position to .5, as shown in Figure 7.20, but do not click OK or Apply yet.

The final step, adding a setback, can also be done with the mini-toolbar, but you will use the dialog box instead.

12. Click the Setbacks tab in the dialog box. Click Vertex 1 at the intersection of the bottom, front, and left faces. Set the Setback values for the three edges to 8.

The standard fillet type will resolve the edges with tangency. There is also an option with edge fillets to apply G2 continuity for smooth transitions.

13. Select Click To Add below Vertex 1 to add another vertex at the intersection of the bottom, front, and right faces. Set their values to 8 as well, as shown in the dialog box in Figure 7.21, and then click OK.

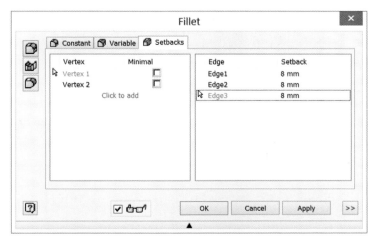

FIGURE 7.21 The Setbacks tab of the Fillet dialog box doesn't offer more options than the mini-toolbar but can be easier to navigate.

Setbacks add a softer corner and are popular on items such as consumer products. By building the constant and variable radius features at once, you can include setbacks.

Adding a Draft Angle

When creating an extrusion, sweep, and a few other features, you can include a draft angle, referred to in the respective dialog boxes as a Taper value. Sometimes it is best to add the draft angle after the fact. With the bottle, for example, you need to include the draft angle on a face that you couldn't add when you created the main feature. For that purpose, you can use the Draft tool.

1. Make certain that the 2014 Essentials project file is active, and then open c07-11.ipt from the Parts\Chapter 07 folder.

2. Be sure that you orbit the part to see the green face on the bottom.

3. Start the Face Draft tool on the Modify panel of the 3D Model tab.

The Face Draft dialog box will prompt you with an icon to select the pull direction. The pull direction is a face or plane around which the draft angle pivots.

4. Select the yellow face on the part to use it for the pull direction.

5. Select the side face of the recessed extrusion near the yellow face, and set the angle to **15 deg**, as shown in Figure 7.22.

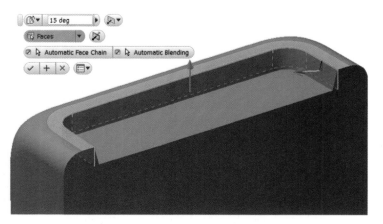

FIGURE 7.22 The draft feature provides an arrow you can drag to change its angle.

Once selected, the preview of the face raft should show the bottom edge of the face moving into the open space. If it doesn't, select the Flip Pull Direction icon to correct it.

6. Click OK to add the draft to the part.

You can also place the draft based on a plane or a sketch. The plane can even be set away from the faces to be drafted.

Replacing One Face with Another

In cases where a complex shape makes up only a portion of a part, you can replace a planar face with a contoured one. To continue building the bottle, you'll use that technique to build a curved face that couldn't readily be created with the original feature.

1. Make certain that the 2014 Essentials project file is active, and then open c07-12.ipt from the Parts\Chapter 07 folder.

At the bottom of the feature tree is a surface feature created by extruding an open profile. It is also possible to revolve and loft a surface by selecting the option.

2. Click the surface that passes through the part.

3. Select the Edit Sketch tool from the screen.
   The model will reposition so the view is normal to the sketch.

4. Press F7 (or right-click and select Slice Graphics) to remove the solid model down to the sketch, as shown in Figure 7.23.

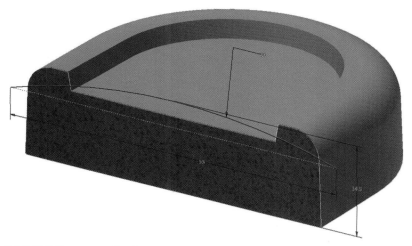

FIGURE 7.23 The Slice Graphics function can expose sketches within the model.

You will notice that there is yellow sketch geometry that makes up the perimeter of the portion of the part that the sketch intersects. This geometry cannot be created using the normal Project Geometry tool. It has to be created using Project Cut Edges.

5. Use Finish Sketch to return to the 3D model.

6. Expand the Surface panel on the 3D Model tab, and select the Replace Face tool.

7. Click the large recessed face on the front of the part.

8. In the dialog box, click the New Faces check box, and then click the surface.

9. Click OK to update the model.

10. In the Browser, right-click the ExtrusionSrf1 feature, and then dese-
    lect Visibility.

Another tool that can be used for this is Sculpt, which you'll use in Chapter 10,
"Creating Sculpted and Multibody Parts." Replace Face is more basic and works
well for substituting geometry.

Fillet Error Handling

Fillet previews make it easier to understand what features will be successful.
When you're dealing with a large number of edges, it can be desirable to use
what will work and clean up what doesn't work — rather than having edges
rejected outright.

In the following steps, you will apply fillets to the edges of the two recesses
on the parts and see what happens if the fillets you apply are not completely
successful:

1. Make certain that the 2014 Essentials project file is active, and then
   open c07-13.ipt from the Parts\Chapter 07 folder.

2. Start the Fillet tool from the marking menu.

3. Set the radius to **1 mm** and select the bottom edges of the two pockets of the part, as shown in Figure 7.24.

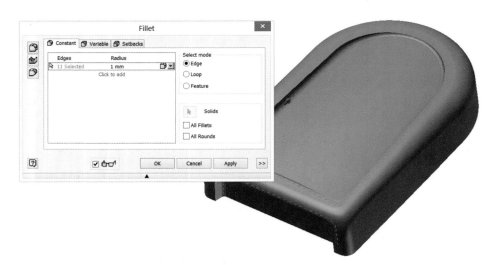

F I G U R E   7 . 2 4 Adding fillets to the pockets of the part

4. Add another fillet size by selecting Click To Add, set its radius to 3 mm, and select the top edges of the pockets.

5. After the preview updates, click OK to generate the fillets.

   An error message will appear (Figure 7.25) informing you that not all of the edges can be filleted at the selected values and offering you an opportunity to accept the successful blends.

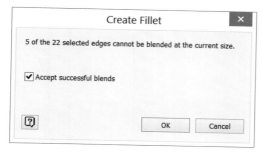

F I G U R E 7 . 2 5 The Create Fillet message

6. Make sure the Accept Successful Blends option is checked and click OK to create the 17 out of 22 edges that can be created.

7. Use the tool to add a 2 mm radius to the edges that failed.

You have been working with half of a component. When components are symmetrical about a plane or an axis (the handle in the earlier exercises), it makes sense to form one part of the component at a time.

With the fillets added and the faces replaced, this side of the part is complete. Next you will start working with the part as a whole.

Mirroring

Pattern tools reduce the tedium of creating repeated features. Creating a mirror feature can save work when there is symmetry in the part, as you'll see next:

Certification Objective

1. Make certain that the 2014 Essentials project file is active, and then open c07-14.ipt from the Parts\Chapter 07 folder.

   The Mirror tool is located in the Pattern panel of the 3D Model tab.

2. Start the Mirror tool.

3. In the Mirror dialog box, select the Mirror A Solid option.

Choosing this option will limit your selection option to clicking the mirror plane.

4. Click the XY Plane under the `Origin` folder, and click OK to create the full body shown in Figure 7.26.

FIGURE 7.26   Use the Mirror tool instead of modeling both sides.

This part could've been modeled as a quarter of the body, mirroring it around the YZ plane and then the XY plane. Once the symmetrical portions are mirrored, you can add unique features to the different areas of the part.

Using a Fillet to Close a Gap

> You can also select individual features rather than mirroring or patterning whole solids.

In Chapter 6, "Exploring Part Modeling," you created a number of work planes and created a feature that was extruded back into a curved face. Sometimes, though, there are benefits to creating geometry in the other direction, but doing so might have consequences too.

If you look closely at the model you are about to open, you will see a gap between the cylinder on the top and the main body. In these situations, applying an edge fillet is not possible because the faces do not intersect. In this exercise you will use a face fillet to bridge this gap.

1. Make certain that the 2014 Essentials project file is active, and then open c07-15.ipt from the Parts\Chapter 07 folder.

2. Start the Fillet tool from the marking menu or the Ribbon.

3. In the mini-toolbar (or dialog box), select the Face Fillet option.

4. Click the cylindrical face of the top boss feature and one of the large curved faces just below it.

    With the default options, you need to click only two faces. Selecting the faces causes the tool to preview a fillet that can solve between the faces.

5. Set the radius to 3, and click OK on the mini-toolbar (Figure 7.27) to generate the fillet.

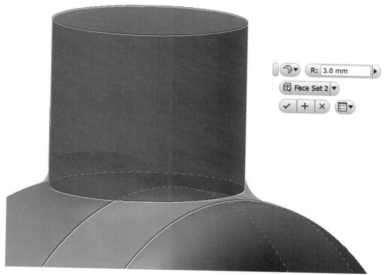

FIGURE 7.27 Creating a fillet between faces can close a gap.

You can use this option to close large gaps or problematic edges when an edge fillet fails.

Adding a Coil

Threaded hole features do not generate physical threads, but if you need threads or want to represent a coil of any type, there is a specific feature for it. In this example, you'll use a coil to create threads for the mouth of the bottle.

Coil

1. Make certain that the 2014 Essentials project file is active, and then open c07-16.ipt from the Parts\Chapter 07 folder.

2. Start the Coil tool from the Create panel on the 3D Model tab.

3. The profile will be automatically selected. Click the y-axis from the Origin folder for the axis of the coil. It might be necessary to use the direction button in the dialog box to make sure the coil is going toward the body of the part.

    This will open the Coil dialog box. This dialog box has three tabs: for selecting geometry, specifying the coil, and controlling the ends of the coil.

4. Switch to the Coil Size tab and set the type to Revolution And Height, set the Height value to **7**, and set the revolution count to **2**, as shown in Figure 7.28.

FIGURE 7.28 Using the Coil tool to generate threads

5. Click OK to generate the coil feature.

This feature can be created using pretty much any profile as long as it doesn't intersect itself. This makes the feature flexible for many uses.

Using Open Profiles

All of the sketched features you've created have used closed profiles to generate solid features and open sketches to create a surface feature. An open profile is capable of creating a solid when it can intersect with existing solid geometry. You'll take advantage of that to create a stop on the neck of the bottle.

1. Make certain that the 2014 Essentials project file is active, and then open c07-17.ipt from the Parts\Chapter 07 folder.

2. Set your view to look at the front of the ViewCube.

3. Start the Revolve tool from the Create panel on the 3D Model tab.

Revolve

4. Click the open profile that is in the sketch. Move the mouse toward the solid until it shows a preview of the area in the sketch that will be filled in. See Figure 7.29.

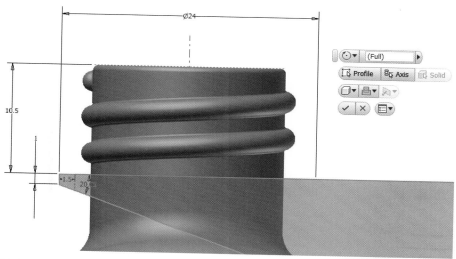

FIGURE 7.29 Previewing a projected profile

5. Click the mouse when the correct area is highlighted to use this profile.

6. Click the y-axis in the Browser for the axis or the centerline in the sketch.

7. Once selected, the preview will change color. Click OK to create the complete feature. Figure 7.30 shows your finished part.

FIGURE 7.30   The finished part showing the revolved lip

Open profiles work well for dealing with parts that include draft features on faces. It is difficult to align a close profile with these faces.

View Representations in a Part

In Chapter 9, "Advanced Assembly and Engineering Tools," you will use view representations to save different configurations for the way an assembly appears. In a part file, you can do the same thing to quickly switch between preselected color or texture options.

1. Make certain that the 2014 Essentials project file is active, and then open c07-18.ipt from the Parts\Chapter 07 folder.

2. Expand the View options in the Browser, and confirm that a view representation called Aqua is the active version.

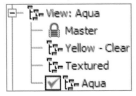

3. Switch between the other saved options to see the options.

If your workflow allows a part to have different colors without having a different part number, view representations can give you the option of placing the part in an assembly using different representations to show different colors.

THE ESSENTIALS AND BEYOND

The advanced filleting tools in conjunction with lofting and sweeping tools open many doors for exploration in part modeling. Patterns and mirrored parts and features are great ways of simplifying and speeding up the creation of parts. Used in combination, these tools make your possibilities endless.

ADDITIONAL EXERCISES

▶ Look at some of your current designs for opportunities to use mirroring and patterning.

▶ Try using fillets instead of sketched curves.

▶ Shell features can be used to create thin parts by removing all but a few faces. Consider whether you have parts on which you could use this technique, and give it a try.

▶ Using work features is a critical skill to master. Practice using them and combining them.

CHAPTER 8

Creating Advanced Drawings and Annotations

In Chapter 4, "Introducing 2D Drawing," you learned how to create the basic drawing views and what some might consider advanced drawing views of a part file. Placing views of assemblies is not any different from placing an individual part file. This chapter expands on those lessons, and you will begin to work with advanced drawing creation and editing tools. You will explore additional dimensioning and annotation tools as well.

▶ **Creating advanced drawing views**

▶ **Using advanced drawing annotation tools**

Creating Advanced Drawing Views

Certification Objective

The part modeling process normally combines the uses of sketched and placed features. In Chapter 4, you created a section view that was based on a sketch. Projected views are placed by projecting from a parent view. Advanced drawing views often use sketches for their definition and can even use more than one sketch. Even a projected view utilizes advanced technologies that help make decisions on your behalf.

Projecting Views from a Section View

You can place as many base views in a drawing as you want, but if you already have a section view with the proper orientation, you can use it as a parent view.

1. Verify that the 2014 Essentials project file is active, and then open the c08-01.idw file from the Drawings\Chapter 08 folder.

2. Right-click the section view, select Projected View from the marking menu, and drag the new view above and to the right of it.

   An isometric view of the geometry that is visible in the section view is created.

3. Click the page to place the view; then right-click and select Create from the context menu to generate the view, as shown in Figure 8.1.

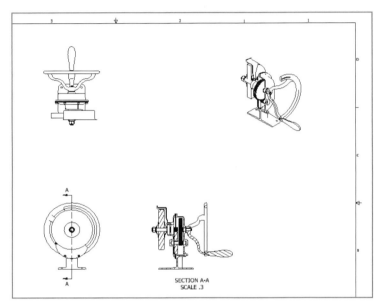

F I G U R E   8 . 1   **Placing an isometric view of a section view**

4. Right-click the section view again, and select the Projected tool from the marking menu.

5. Place the view to the right of the section view, and create the new view. Figure 8.2 shows the result.

The new view is not a partial view; it is a full view. Because the new view is an orthographic projection, all of the geometry is included.

Creating a Sketch in the Drawing View

There are many uses for creating sketches in a drawing view. In this case, you need to create geometry to modify the drawing view:

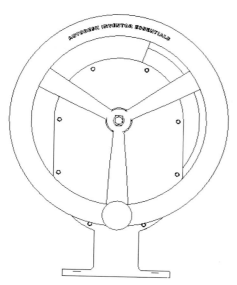

FIGURE 8.2 A complete view can be projected from a section view.

1. Verify that the 2014 Essentials project file is active, and then open c08-02.idw from the Drawings\Chapter 08 folder.

2. When you hover over the bottom-left view, the boundary will highlight. When the view highlights, click the view, and then select Create Sketch in the Sketch panel of the Place Views tab.

3. When the Sketch tab appears, draw a spline in the view similar to the one in Figure 8.3.

4. After the loop of the spline is closed, right-click and select Finish Sketch on the marking menu to finish the sketch.

    It is important to verify that the sketch is associated with the drawing view.

5. To verify association, drag the drawing view, and make sure the sketch moves with it.

If the sketch is not attached to the drawing view, delete the sketch you made, and go through the steps again to make sure that the view is highlighted when you start the Create Sketch tool.

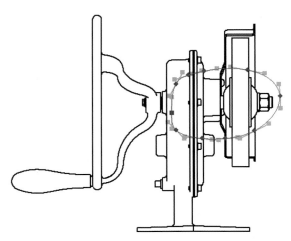

FIGURE 8.3 Define a sketch in the drawing view to make changes to the view.

Generating the Break Out View

The sketch you created will be used as a boundary to remove part of the components of the assembly so you can see the interior.

1. Verify that the 2014 Essentials project file is active, and then open c08-03.idw from the Drawings\Chapter 08 folder.

Break Out

2. Click the Break Out tool in the Modify panel of the Place Views tab.

3. Select the left view, and when the Break Out dialog box opens, you will notice it has already selected the spline since it was the only closed profile.

   You will then need to select the depth to which you cut through the parts.

4. Hover over a curved edge. When the green dot appears, click it, as shown in Figure 8.4.

5. Click OK to generate the view.

There are several options for selecting what components will be cut using the Break Out tool.

This generates and cuts any and all parts down to the selected point, but the traditional line view drawing still makes it difficult to distinguish the components.

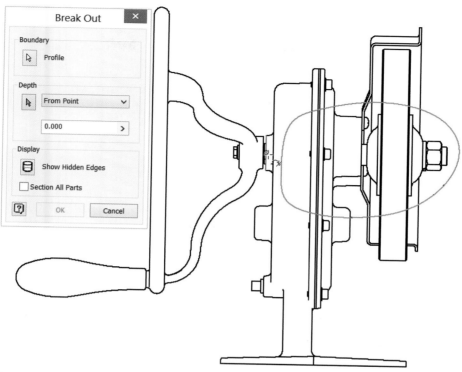

FIGURE 8.4 After you select the profile, you determine the depth of the cut.

Changing Part Drawing Behavior

In this exercise, you will modify how the Autodesk® Inventor® software processes generating the drawing view of certain parts in the assembly:

1. Verify that the 2014 Essentials project file is active, and then open c08-04.idw from the Drawings\Chapter 08 folder.

2. Double-click the break out view to edit it.

3. Deselect the Style From Base check box.

4. Select the Shaded icon, and click OK to update the view.
   In Figure 8.5, you can see there are a couple of errors. The first is that the screw and washer appear to be floating in the middle of the view. The second is a gear that is sectioned because of the way it was created. The third item flagged is not a mistake, but you want to show the seal on the shaft sectioned.

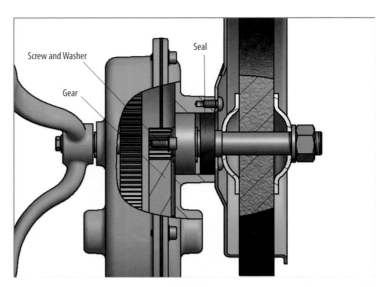

FIGURE 8.5 You can control whether parts are sectioned or even visible in a drawing view.

When you move your cursor over the view, all the edges highlight as you pass over them. By default, Inventor uses an Edge priority for selecting geometry in a drawing. You want Inventor to highlight the parts as you pass over them instead. This will make selecting them much easier.

5. Using the Selection drop-down on the Quick Access toolbar, change the current selection priority to Part Priority by selecting it from the list.

6. Hold the Ctrl key to select the screw and the washer. Right-click the edge of one of the parts, and then select Section Participation from the context menu.

7. When the menu expands, select Section.
    The view regenerates with the screw and washer affected by the break out section, and therefore, they disappear.

8. Select the gear, and choose None from the Section Participation options.

9. Select the seal, and then choose Section.
    This will again update the view to show the seal sectioned around the shaft. These changes will each update the view to appear like Figure 8.6.

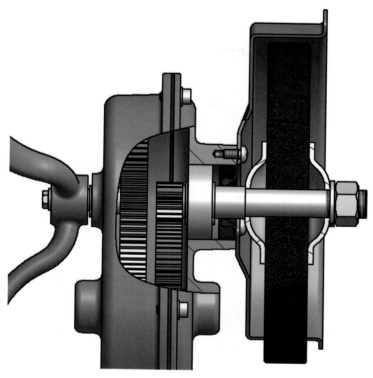

F I G U R E 8 . 6 The updated view with the updated part visibility options

Removing Part Visibility

If the screw and washer in the previous exercise weren't in a section or break out view, it would be possible to just remove the visibility. To do this, you must break the association between the drawing view and the design view of the assembly using the Drawing View dialog box. This has been done for you in the file referenced in the next exercise.

Certification
Objective

1. Verify that the 2014 Essentials project file is active, and then open c08-05.idw from the Drawings\Chapter 08 folder.

2. From the Quick Access toolbar, set the selection filter to Part Priority.

3. Hover over the shroud. When it highlights, right-click and deselect Visibility.

The drawing view regenerates the previously hidden edges that are now visible (Figure 8.7).

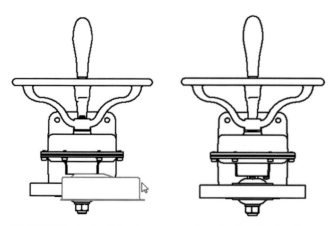

FIGURE 8.7 Selecting a part and turning it off in the view

Implementing View Suppression

A drawing view can be a stepping-stone to generate another, more important view. Being able to remove the clutter of the parent view can be useful.

1. Verify that the 2014 Essentials project file is active, and then open c08-06.idw from the Drawings\Chapter 08 folder.

   View1 is the base view in the drawing. It is located in the upper-left corner.

2. Find View1:c08-09.ipt in the Browser, hover over it, and when the view highlights, right-click.

3. Select Suppress from the context menu.

Even though View1 is the base view for all the other views, you removed it without disruption to the rest of the drawing.

Drawing Element Suppression

You can suppress individual lines and other elements of the drawing view:

1. Verify that the 2014 Essentials project file is active, and then open c08-07.idw from the Drawings\Chapter 08 folder.

2. Zoom in on the isometric view in the upper-right corner.

3. Click and drag a window from left to right around the entire view.

4. When the lines in the drawing highlight, right-click and deselect Visibility from the context menu. Figure 8.8 shows the result.

FIGURE 8.8 A view can have visible
edges removed for a more realistic look.

Once an edge has been suppressed, it will remain that way even if the geometry changes size. New geometry added to the components will appear in the drawing view, so it is good practice to monitor suppressed edges.

TIP   You can restore edges that you've made invisible by right-clicking the view and selecting Show Hidden edges from the context menu. The hidden edges will highlight, and you can pick the ones you want visible again or even Show All. When you're finished, select Done from the context menu.

Using Break View

When a part is long or has a large portion that consists of the same geometry, a *break* view can remove a portion of the drawing to make it easier to detail the dissimilar portions.

1. Verify that the 2014 Essentials project file is active, and then open c08-08.idw from the Drawings\Chapter 08 folder.

2. Zoom out until you can see the entire drawing view as it extends beyond the border of the drawing.

Break

3. On the Place Views tab, select the Break tool in the Modify panel.

4. Click the Drawing view to open the Break dialog box.

5. Set the Gap value to .5.

6. Click the first point of the break just to the right of the red part and the second just to the left of the threaded end, as shown in Figure 8.9.

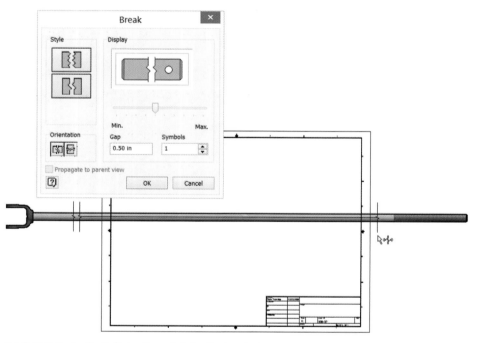

FIGURE 8.9 Defining the points for the break view

7. Add a dimension to the length of the gray shaft (Figure 8.10).

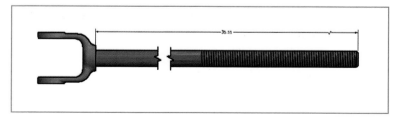

FIGURE 8.10 Adding a dimension to the break view

The dimension will show the true length of the visible portion of the bar.

Creating a Slice View

A *slice* view is different from a section view in that only the geometry of the cut portion is visible. It can be used for something like a rotated section or to modify another view at cutting planes using a sketch.

1. Verify that the 2014 Essentials project file is active, and then open c08-09.idw from the Drawings\Chapter 08 folder.

2. Select the Slice tool in the Modify panel of the Place Views tab.

3. Click the isometric view as the view to be modified, and in the dialog box select the Slice The Whole Part check box.
   Once the target view is selected, you need to select a sketch that will be used to cut the view.

4. Select the vertical lines in the side view of the part, and click OK. Figure 8.11 shows the result.

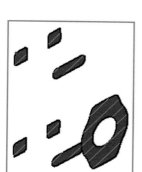

FIGURE 8.11 A slice view can show how the shape of a component is structured.

This updates the isometric view to show the portions of the part where the lines pass through it.

Placing a Custom View

The ViewCube® in parts and assemblies aligns with the standard view options for placing drawing views. At times, these views just can't properly describe the geometry. For these times, you can create a *custom* view.

Base

1. Start a new drawing using the ANSI (mm).idw template available in the Metric folder.

2. Select the Base tool in the Create panel of the Place Views tab or from the marking menu.

3. When the Drawing View dialog box opens, use the Browse tool to locate the files. Find c08-01.ipn in the Assemblies\Chapter 08 folder.

    The .ipn extension is for Inventor presentation files. A presentation file is used to create an exploded view of an assembly.

4. Once the exploded assembly is selected, its preview appears. Below the list of standard view orientations is the Change View Orientation button. Click it.

    This opens the presentation file in a specialized Custom View tab with file-viewing tools.

5. Position the exploded view similar to Figure 8.12.

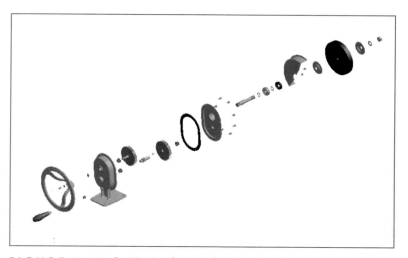

FIGURE 8.12   Positioning the parts for a drawing view

6. Once the view is positioned, click the Finish Custom View tool at the end of the Custom View tab.

7. The preview updates to the view you established. Set the scale so the exploded view of the assembly fits across the page, as shown in Figure 8.13. Make it a shaded view.

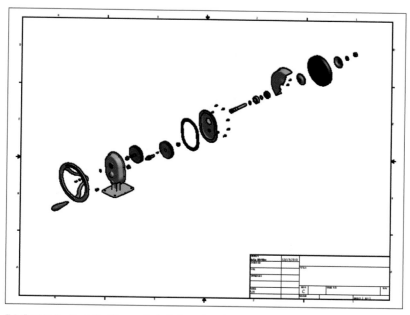

FIGURE 8.13 The exploded view placed in the drawing

8. Click the location for the view on the drawing sheet, and then right-click and select OK to finish the command and place the view.

The presentation file can also create animated assembly instructions, and you will do that in Bonus Chapter 4, "Creating Images and Animations from Your Design Data."

 TIP   A nice but less commonly used capability is to copy drawing views on the same page. This approach can be handy if you need to duplicate views to distribute detailing or want to save time setting up an alternative presentation of the views.

Using Advanced Drawing Annotation Tools

Chapter 4 explored the elements of dimensioning that will make up the bulk of typical detailing. Many other tools are necessary to make a complete drawing. The next group of exercises will focus on these tools.

Certification
Objective

Using Automated Text

Automated text tools make it possible to add information from a number of different model properties to title blocks and annotation. In this exercise you will add a note that will include the mass of the part and that will update automatically.

1. Verify that the 2014 Essentials project file is active, and then open c08-10.idw from the Drawings\Chapter 08 folder.

2. Just above the title block is a line of text. Double-click the text to open the Format Text dialog box.

3. In the Format Text dialog box, change the Type drop-down to Physical Properties – Model and the Property drop-down to Mass.

4. Click the Add Text Parameters button to add the model property to the text-editing window below.

5. Click OK to update the text.
   The value of the mass appears, but it will likely be listed as N/A because the part's Mass properties are out-of-date. To update them, you need to open the part file and update it.

6. Right-click one of the drawing views on the drawing sheet or in the Browser, and select Open from the context menu.

7. When the part file opens, right-click the Part icon at the top of the Browser and select iProperties. This opens the c08-09.ipt iProperties dialog box.

8. Click the Physical tab, and click the Update button.

9. Return to the drawing to see the results.

> THE APPROXIMATE WEIGHT OF THE FINISHED PART IS 0.237 lbmass

10. Close both the drawing and the part file without saving the changes.

Placing Leader Text

You can use the Text tool to place text in a rectangular space on the drawing. The Leader Text tool automatically places the text at the end of a leader.

1. Verify that the 2014 Essentials project file is active, and then open c08-11.idw from the Drawings\Chapter 08 folder.

2. You can find the Leader Text tool on the marking menu. You can also find it in the Text panel on the Annotate tab. Start the Leader Text tool.

3. Click the top machined surface of the part, and then click a second point to locate where the text will be generated.
   You can continue to click points to string the leader through.

4. Right-click and select Continue from the context menu to open the Format Text dialog box.

5. Type in anything you like, and then click OK to generate the text. See Figure 8.14 for an example.

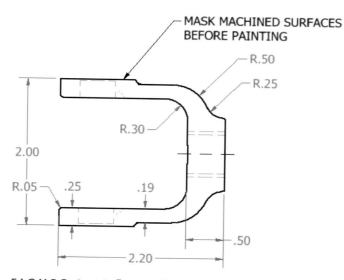

FIGURE 8.14 Text attached to the leader

 TIP Notice that you can create multiple rows of text in the leader.

Including Special Symbols

Many types of symbols are needed for drawings. Inventor includes a wide selection of symbols and the ability to define your own. These symbols can be placed directly or attached to a leader.

Surface

1. Verify that the 2014 Essentials project file is active, and then open c08-12.idw from the Drawings\Chapter 08 folder.

2. In the Symbols panel of the Annotate tab, select the Surface tool.

3. Click the top surface that the leader text is attached to.

4. Right-click and select Continue from the context menu to open the Surface Texture dialog box (Figure 8.15).

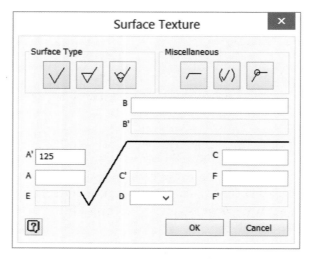

FIGURE 8.15 The Surface Texture dialog box is very thorough.

5. Make any changes in the dialog box, and click OK to place the symbol.

6. The Symbol tool will continue to work. Click the face on the far right, drag a leader away from the part, and click a second point for the leader.

7. Right-click and select Continue to launch the dialog box again, and click OK to place a Surface symbol on the leader. See Figure 8.16.

8. Right-click and choose Cancel, or press the Esc key to stop the Surface symbol tool.

Although this exercise covers only one type of symbol, nearly all of them work in the same way.

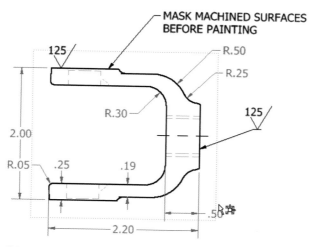

FIGURE 8.16 Symbols can be attached to leaders as well.

Creating a Balloon Callout

The exploded presentation view is a great way to display the physical makeup of an assembly, but you also need to be able to document what the components are. Calling out the parts using *balloons* and generating a list of the parts can help create a more complete document.

Certification
Objective

1. Verify that the 2014 Essentials project file is active, and then open c08-13.idw from the Drawings\Chapter 08 folder.

2. The Balloon tool is on the marking menu (in an open part of the Design window) and on the Annotate tab in the Table panel. Start the tool.

3. The Balloon tool opens in component-selection mode. Click the wheel on the left end.

    This launches the BOM Properties dialog box, where you can define the source of the balloon information based on the bill of materials (BOM) structure.

4. Leave the BOM View value as Structured, and click OK.

    As you place a balloon, it moves smoothly but snaps at 15-degree increments. Before the position of the leader is set, you can override the snap. Hold the Ctrl key as you position the balloon, and it can be placed at any angle.

5. Click a second point to set the length of the leader.

6. Right-click and select Continue to place the balloon. Then, press the Esc key to end the Balloon tool. See Figure 8.17.

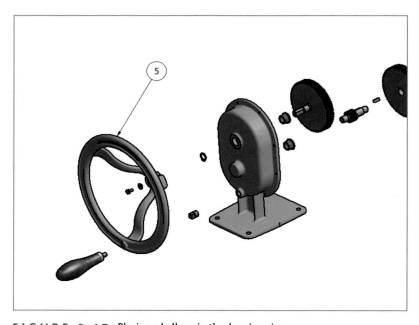

FIGURE 8.17 Placing a balloon in the drawing view

7. Click the leader of the balloon, and the grip points will highlight. Click the point at the tip of the arrow, and drag it to the housing to the right.

8. When the edge of the part highlights, release the mouse to attach the balloon to the other part.

The balloon number will update to correspond with the value gathered from the BOM contained in the assembly.

Using the Auto Balloon tool

The Balloon tool is effective, but a faster alternative is Auto Balloon.

1. Verify that the 2014 Essentials project file is active, and then open c08-14.idw from the Drawings\Chapter 08 folder.

2. The Auto Balloon tool is on the Annotate tab in the Table panel. You will have to expand the Balloon tool flyout to access it. Start the Auto Balloon tool.

   This tool works a little differently; it opens a dialog box (see Figure 8.18) immediately. The dialog box has a lot of options for determining how to prioritize what information is entered into the balloons.

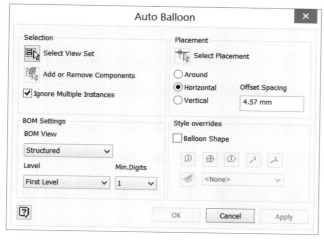

FIGURE 8.18 The Auto Balloon dialog box

3. Click the drawing view if the Select View Set button is depressed (see Figure 8.18). This activates the Add Or Remove Components tool.

4. Drag a window around several components toward the right end of the exploded view. See Figure 8.19.

5. In the Placement group of the dialog box, set Offset Spacing to **10**. Then click the Select Placement icon, and move your cursor over the screen to see a preview of the balloons.

6. Click to place the balloons on the screen, and then click OK.

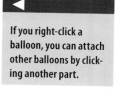

If you right-click a balloon, you can attach other balloons by clicking another part.

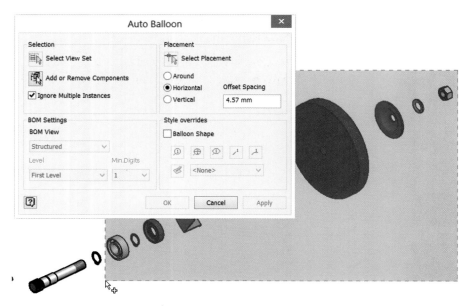

FIGURE 8.19 Select the components to be ballooned.

Creating a Parts List

Certification
Objective

The *parts list* is a way of documenting the BOM of the assembly. You do not have to present the entire BOM and are not limited to items contained in it.

1. Verify that the 2014 Essentials project file is active, and then open c08-15.idw from the Drawings\Chapter 08 folder.

2. In the Table panel of the Annotate tab, start the Parts List tool.

3. The Parts List dialog box that opens (Figure 8.20) has a number of options similar to those in the Auto Balloon dialog box.

4. For simplicity, you can leave the options alone. Click the drawing view, and then click OK.

5. Place the parts list in the drawing.

6. Drag the left edge of the column to resize the various columns (Figure 8.21).

7. Moving over portions of the text presents a cursor that allows you to click and drag the location of the parts list.

8. Locate the parts list above the title block and against the right border of the drawing. The parts list will snap into place. See Figure 8.22.

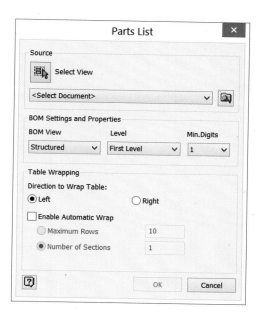

FIGURE 8.20 The Parts List dialog box has many options.

ITEM	QTY	PART NU
1	1	Grinder Base
2	1	Grinder Face
3	1	Spur Gear2
4	1	Gasket
5	1	Flywheel Handle

FIGURE 8.21 Drag the separation lines to resize the parts list's columns.

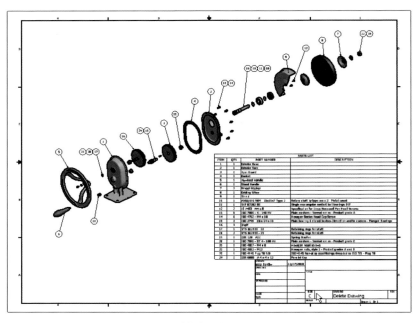

FIGURE 8.22   The parts list added to the drawing

> **TIP**   The parts list is based on the structure of the BOM, and that structure can be represented in the parts list as well. The parts list would show a parent component and the tree of its child components. Additional properties can be added to the parts list.

The initial size of the columns and other properties of the parts list are controlled by the style and standards. This is another element that will likely be set once within a company and left alone. A parts list can also be split into separate portions that can be repositioned to position them near specific callouts.

Editing Dimension Values

When you set the dimension styles, the purpose was to make as many dimensions as possible appropriate for your needs. That doesn't mean you won't need to make changes to some.

1. Verify that the 2014 Essentials project file is active, and then open c08-16.idw from the Drawings\Chapter 08 folder.

2. Double-click the 66.00 dimension. This opens the Edit Dimension dialog box.

3. Choose the Precision And Tolerance tab, and set Tolerance Method to Symmetric and the Upper value to **0.01**, as shown in Figure 8.23.

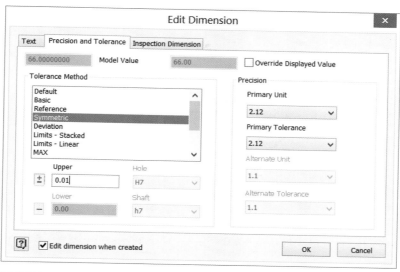

FIGURE 8.23   The Edit Dimension dialog box

4. Click OK to update the dimension, as shown in Figure 8.24.

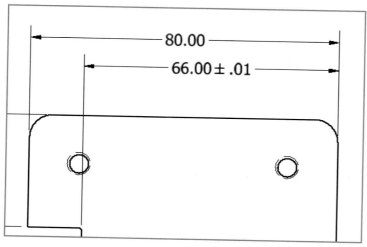

FIGURE 8.24   The dimension updated with the tolerance

There are also other ways to dimension the part.

Placing Ordinate Dimensions and Automated Centerlines

Ordinate dimensions (like baseline dimensions) can be placed either as individuals or as a set. The Ordinate Set option keeps the members styled as a set and also automatically does some spacing of the dimensions.

1. Verify that the 2014 Essentials project file is active, and then open c08-17.idw from the Drawings\Chapter 08 folder.

2. Right-click within the boundary of the view, and select Automated Centerlines from the context menu.

   The Automated Centerlines dialog box (Figure 8.25) allows you to select whether to include cylindrical extrusions or faces that have a radius, and you can also choose whether to select objects with an axis that is normal or parallel to the view.

3. Keep the defaults, and click OK to place the centerlines.

4. Locate the Ordinate dimension tool in the Dimension panel on the Annotate tab, expand it, and select the Ordinate Set tool.

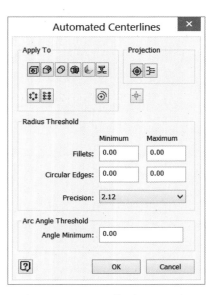

FIGURE 8.25 The Automated Centerlines dialog box

5. In the status bar, notice the prompt to click an origin point. Click the center mark of the large hole in the lower-right corner of the part.

6. Click the vertical edges of the part shown in red in Figure 8.26, and the center marks highlighted with green dots.

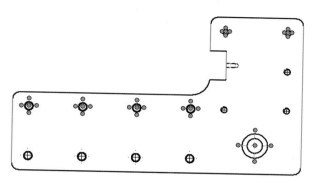

FIGURE 8.26 Selecting elements of the drawing view on which to place ordinate dimensions

7. When you've selected all the elements, right-click and select Continue from the context menu.

8. Drag the location of the dimensions above the model geometry, and click to place the dimensions.

9. Right-click and select Create to complete the dimensions. Figure 8.27 shows the result.

Ordinate dimensions can be placed in more than one direction on the view.

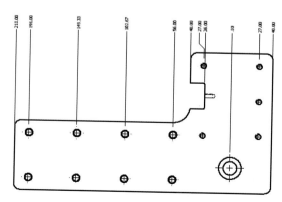

FIGURE 8.27 The ordinate dimensions placed in the drawing

Hole Tables

When you have a large number of holes, it is more efficient to use a hole table than to dimension the location of each one.

1. Verify that the 2014 Essentials project file is active, and then open c08-18.idw from the Drawings\Chapter 08 folder.

2. The Hole Table flyout is located in the Table panel on the Annotate tab. Expand the list, and click the Hole View option.

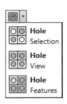

3. Select the drawing view of the part.

4. Set the origin location to be the center of the large hole in the lower right of the part.

5. When the preview of the table appears, drag it near the top of the title block against the left edge of the border, and click to place it.

   Closer inspection of the drawing will show that it has labeled each of the holes with an alphabetical hole class and then a number that counts how many holes of that size are in the view. The table shows those hole names along with the dimensional information for placing the holes, as well as the description of each hole. To simplify the table, you can group the descriptions together.

6. Double-click the hole table to open the Edit Hole Table: View Type dialog box. Select the Combine Notes radio button (Figure 8.28) in the Row Merge Options group.

7. Click OK to update the table. Figure 8.29 shows the result.

For many CAD users, creating detail drawings isn't just a final step; it is the most critical. Many companies use these drawings not only to produce their products but as the legal documentation of their products. It takes a strong tool set to produce them accurately and efficiently. Inventor has these tools.

There are still more types of annotation that are more specialized, and I will cover them along with the design techniques they document in later chapters.

Edit Hole Table: View Type

| Formatting | Options |

Row Merge Options

○ None
☑ Reformat Table on Custom Hole Match
☐ Numbering

○ Rollup
☐ Delete Tags on Rollup
☐ Secondary Tag Modifier on Rollup

◉ Combine Notes
☑ Reformat Table on Custom Hole Match
☐ Numbering

Hole Tag Options
☑ Preserve Tagging

Tag Order
◉ Arrange by Position
○ Arrange by Size
☑ Group Hole Types

View Filters

Included Features
☑ ⊚   ☐ ⊹
☑ ⌂   ☐ ⊞

Included Hole Types
☑ ⊔   ☑ ⊽
☑ ⧅   ☑ ⊽

OK   Cancel   Apply

FIGURE 8.28 Merging hole descriptions can make the table less confusing.

HOLE TABLE			
HOLE	XDIM	YDIM	DESCRIPTION
A1	-196.00	-12.50	
A2	-149.33	-12.50	
A3	-102.67	-12.50	
A4	-56.00	-12.50	Ø6.60 THRU
A5	-196.00	32.50	
A6	-149.33	32.50	
A7	-102.67	32.50	
A8	-56.00	32.50	
B1	-27.00	32.00	
B2	27.00	32.00	
B3	27.00	67.00	M6x1 - 6H
B4	-27.00	102.00	
B5	27.00	102.00	
C1	.00	.00	Ø13.49 THRU ⌴ Ø22.23

FIGURE 8.29 The revised hole table showing the combined notes

THE ESSENTIALS AND BEYOND

The ability to manipulate the appearance of components within an assembly is also helpful for creating effective documentation. Assembly oriented annotation tools like Balloons and Parts Lists are subject to personal and company standards. These tools are flexible enough to allow you to tailor them to your needs. Once you have made modifications to special annotations be sure to create templates from your work to avoid having to re-create it in the future. Review Bonus Chapter 1 for more information on personalizing the drawing styles and standards to your needs.

ADDITIONAL EXERCISES

▶ Experiment with what you learned in this chapter to develop an appropriate parts list style.

▶ Work with the Text tool to see how easy it is to build stand-alone notes.

▶ Explore and compare the behavior of the Ordinate and Ordinate Set tools.

▶ Test the various Tolerance and Fit callouts available in the Edit Dimension tool.

CHAPTER 9

Advanced Assembly and Engineering Tools

Most of the tools needed for the day-to-day work of building assemblies were covered in Chapter 5, "Introducing Assembly Modeling." In this chapter, you will use a few more tools to build your assembly.

Being able to navigate in the assembly and maintain your system's performance is also important, and large assemblies can put you at risk of losing data if your system runs out of memory. Autodesk® Inventor® software representations can solve these problems, and you will see how to control your assembly with them.

Much of the focus of this chapter will be on tools that allow you to work more efficiently by creating geometry in the context of the assembly. These tools also offer a unique opportunity by incorporating engineering calculators. You will see how to create common machine components, but you will also see how to make sure they will do the job.

▶ **Controlling the assembly environment**

▶ **Using design accelerators**

▶ **Working with additional assembly tools**

Controlling the Assembly Environment

An often-overlooked aspect of working with assemblies is controlling both the look and feel of the assembly's view representation and, most important, its system demands. Both of these elements are easy to control but are not automatic. There are three kinds of *representation* in Inventor that enable this enhanced level of control: view, level of detail, and positional. In this section, you'll explore the first two.

Creating View Representations

View representations allow the user to control the visibility and appearance of components in the assembly. Creating several view representations allows a user or multiple users to see the assembly in whatever way they want.

When you close a file the active view will be automatically saved as a View Representation. You can change this by right-clicking the view representation, choosing Save Current Camera, and deselecting Autosave Camera. While editing, you can also restore the saved camera view.

1. Make certain that the 2014 Essentials project file is active, and then open the c09-01.iam file from the Assemblies\Chapter 09 folder at this book's web page, www.sybex.com/go/inventor2014essentials.

2. Locate the Representations folder in the Browser. Expand the folder, and expand the View group.

3. Right-click the Default view representation, and select Copy from the context menu.

   This creates a new view representation named Default1 at the bottom of the list.

4. Click the new view representation twice slowly, and rename it **Natural Colors**.

5. Double-click the Browser icon of the new view representation to activate it.

6. Right-click the Natural Colors representation, and select Remove Appearance Overrides from the context menu.

7. Select c09–01:1 in the Browser, right-click, and then deselect Visibility in the context menu. Figure 9.1 shows the results.

FIGURE 9.1 View representations can save variations of color and visibility.

8. Double-click the icon next to Default to set it as the active view representation, and then switch the view representation back.

The updates to the assembly affect only the visibility and colors of the components in the assembly. Another form of representation has a stronger effect.

Creating Level-of-Detail Representations

Level-of-detail representations will not just remove a component from visibility; they will remove it from memory. If you need to work on large assemblies with limited system resources, the level-of-detail (LOD) representation is an invaluable tool.

Certification Objective

1. Make certain that the 2014 Essentials project file is active, and then open the c09-02.iam file from the Assemblies\Chapter 09 folder.

2. Locate the Representations folder in the Browser. Expand the folder, and expand the Level Of Detail group.
   On the right end of the status bar, two sets of numbers report the number of components displayed in the Design window and the number of documents being accessed by Inventor.

3. Double-click the icon next to the Gears representation.
   This reduces the number of displayed component occurrences and the number of open documents by removing all the components other than the shafts and gears that drive the mechanism.

4. Activate the All Content Center Suppressed representation and see the effect.

5. Right-click the All Content Center Suppressed LOD, and select Copy from the context menu.

6. Rename it as **LOD Primary Components.**

7. Double-click the Browser icon of the Primary Components LOD to activate it.

8. In the Design window, click the wooden handle.

9. Right-click, expand the Selection options in the context menu, and select Component Size as the selection filter for the assembly.
   A dialog box opens that shows the volume of the selected part.

A substitute LOD can replace the assembly with a single part or a shrink-wrapped version of the assembly to improve performance while remaining associative to the assembly. Bonus Chapter 2, "Working with Non-Inventor Data," has lessons on the Shrinkwrap tool.

By default, Inventor will load assemblies with more than 500 unique components in Express mode, which will increase graphics performance on very large files. Express Mode can be disabled in the Application Options ➤ Assembly tab.

10. Change the selection value to **205**, and click the check mark to select the components smaller than 205 millimeters.

11. Right-click and select Suppress from the context menu. The effects are shown in Figure 9.2.

FIGURE 9.2 LOD representations can remove components from memory.

Certification Objective

You must save your work to preserve the changes made to an LOD representation before switching to another one.

POSITIONAL REPRESENTATIONS

Positional representations can override whether a constraint is enabled or suppressed. They can also use different offset values for a constraint. For example, a positional representation named Open can change the value of a Mate constraint from 0 mm to 10 mm, which will cause a gap to appear between parts. These representations can be used to create overlays in a drawing to show an alternative position or be used in an animation (see Bonus Chapter 4, "Creating Images and Animation from Your Design Data") to reduce the number of steps needed to create the animation.

Using Design Accelerators

Inventor has a number of specialized tools that build common machine components. Design accelerators come in two basic classes: calculators that work out
what geometry you will need (such as weld sizes) and generators that can simply
create components or use their built-in calculation tools to help you select the
appropriate components, such as bolts or the size of a shaft.

Certification
Objective

It helps to build something when learning advanced tools. In this chapter, you
will be working on the design of a hand-driven metal grinder.

Using the Bearing Generator

Many bearing standards are global, so this tool allows you to choose from a
number of components with the same basic geometry.

1. Make certain that the 2014 Essentials project file is active, and then
   open the c09-03.iam file from the Assemblies\Chapter 09 folder.

2. Click the Bearing tool in the Power Transmission panel of the
   Design tab.

   The Bearing Generator dialog box that opens has two tabs. The
   Design tab displays a bearing (selected based on the size specified),
   and the Calculation tab allows you to verify that the bearing will
   hold up to the demands of its job. All generators have at least these
   two tabs.

3. Set the dimensional values for the dialog box to be the same as
   Figure 9.3, and click the Update button to build a list of bearings that
   meet the physical specs.

4. Click the down arrow near the top of the dialog box where it says
   Angular Contact Ball Bearings to open a selection dialog box.

5. In the dialog box, use the drop-down menu in the upper right to
   change the category of bearing to Cylindrical Roller Bearings, and
   then select the Cylindrical Roller Bearings (Figure 9.4) option in the
   upper left so that it creates a list of all of the bearings that fit.

6. Scroll down the list, select the SKF NJ bearing, and click OK to create the geometry. You will have to click OK in the File Naming dialog
   box as well.

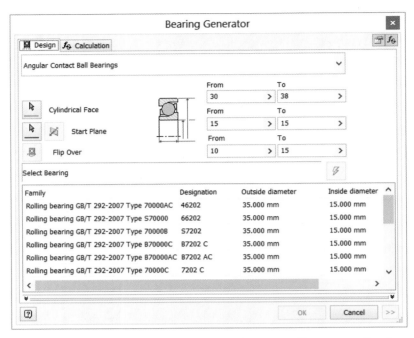

FIGURE 9.3 The Bearing Generator dialog box can quickly generate a list of bearings based on size.

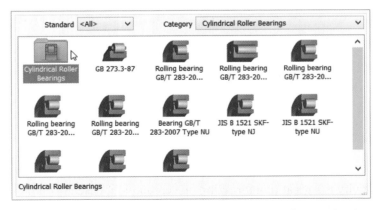

FIGURE 9.4 You can change the category of bearing that you place in the assembly.

7. Click to place the bearing off to the side of the cover.

Using the Calculation tab, you can also test the bearing by checking how the bearing meets life span, loading, and speed ratings to be sure that it will hold up.

Using the Adaptivity Feature within the Assembly

Adaptivity is not a function of a design accelerator, but it is an advanced capability unique to Inventor. It allows part features to be modified by assembly constraints in certain conditions.

Certification
Objective

It is possible to identify an adaptive item in the Browser by locating the red-and-blue icon next to the part or feature or sketch.

1. Make certain that the 2014 Essentials project file is active, and then open the c09-04.iam file from the Assemblies\Chapter 09 folder.

2. Switch the Ribbon to the Inspect tab, and start the Distance option of the Measure tool from the Measure panel, press the M key, or select Measure Distance from the marking menu.

3. Click the inside edge of the large hole in the tan part.
   The hole in the part will measure 38 mm. The bearing in the assembly has an outer diameter of 35 mm. Adaptivity will allow the diameter of the hole to be changed to fit the bearing.

4. Start the Constrain tool in the Relationships panel of the Assemble tab.

5. Hover over the exterior cylindrical face of the bearing.

6. When the axis of the bearing highlights, right-click and click Select Other from the context menu.

7. Click the first Face option from the list that highlights the outer diameter of the bearing, as shown in Figure 9.5.

8. Select the inside face of the large hole in the front cover, and click OK.

9. Click the front face on the ViewCube® so you can see that the diameter of the hole is matched to the diameter of the bearing.

Adaptivity is a great tool for sizing one component based on another, but using it on too many features or parts can slow your performance. It is best used to establish a size, and then it can be disabled to lock that size in.

By default components placed by design accelerators might not relocate if geometry they are aligned with is changed in the assembly. You change this option by right-clicking the component in the Browser, opening the Component flyout, and selecting Automatic Solve.

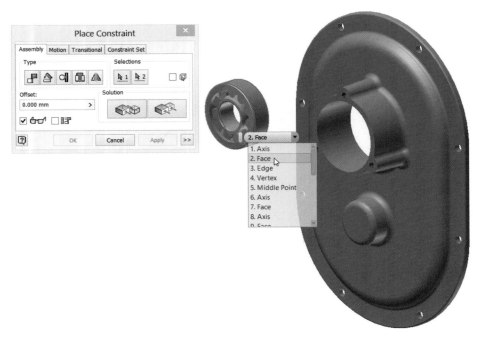

FIGURE 9.5   Use the Select Other option to click additional selection options for placing assembly constraints.

Using the Shaft Component Generator

It is a simple process to create a shaft by revolving a half section or creating a series of extrusions stacked end to end. The Shaft Component Generator design accelerator constructs the shaft through a dialog box and lays out the segments of the shaft.

1. Make certain that the 2014 Essentials project file is active, and then open the c09-05.iam file from the Assemblies\Chapter 09 folder.

Shaft

2. Hold down your Ctrl key and click the Shaft tool in the Power Transmission panel of the Design tab.

   Holding the Ctrl key is necessary only if you want to start a design accelerator with the default options.

   When the dialog box shown in Figure 9.6 opens, shaft sections appear in the large window. Each segment row has columns for the left and right end treatments, the section shape, and any special additions. The size and shape are listed to the right as well.

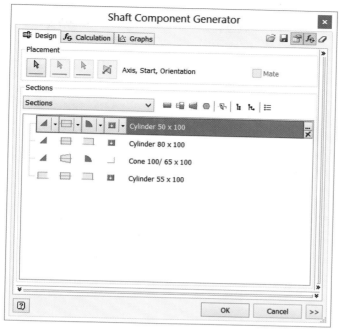

FIGURE 9.6 The Shaft Component Generator dialog box showing the default shaft segments

3. Select the top segment. At the far right, click the Delete icon, and then click Yes in the dialog box that appears.

4. Delete all but the segment listed as Cylinder 55 × 100.

5. Double-click the description of this segment.

6. Change the D (diameter) and L (length) values to 10 mm by clicking the current values in the Cylinder dialog box that appears.

7. Click OK to update the size of the segment.

8. Locate and select the Insert Cylinder icon just above the updated section. This will insert a duplicate segment below the original.

9. Edit the size of the new segment to D = 12, L = 3, and click OK.

10. Click the Insert Cylinder tool again, and create a new segment with D = 25 and L = 12 values.

11. Create another new cylinder, setting D = 12 and L = 16.

12. Create the fifth segment to be the same as the first: D = 10 and L = 10.

13. In the bottom segment, click the end treatment icon on the right, and click the Chamfer option in the drop-down.

14. When the small dialog box appears, set the Distance value of the chamfer to .3, and click the green check mark.

15. Add a .3 chamfer to the left end of the first shaft segment. The Shaft Component Generator dialog box now should look like Figure 9.7.

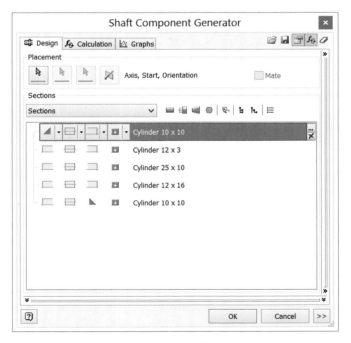

FIGURE 9.7 Adding chamfers to the ends of the shaft

16. Click OK to generate the shaft and click OK again to accept the names of the new files.

17. Place the new shaft into the assembly by clicking a location in the Design window.

At this point, you've developed the geometry that fits in the assembly and that you think will handle the loads. Part of every generator in the Power Transmission tools can also calculate whether the geometry is up to the task at hand. Now, you should see whether this shaft will work.

Calculating and Graphing Shaft Characteristics

For this exercise, you will continue using the shaft you are in the middle of defining. The Shaft Component Generator's Calculation and Graphs tabs have nothing to offer until you've defined geometry on the Design tab. In turn, the Graph tab cannot supply information until the calculation is done.

The Shaft Component Generator is great for hubs, too. A drop-down in the Sections group allows you to add internal geometry from either end.

1. Right-click the shaft you placed in the assembly, and select Edit Using Design Accelerator from the context menu.

2. Click the Calculation tab (Figure 9.8) to see the available options.

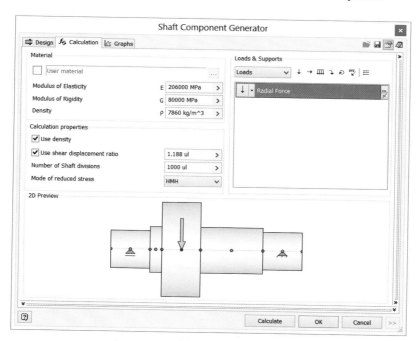

FIGURE 9.8 The Calculation tab of the Shaft Component Generator

The Calculation tab is where you can override and experiment with the material of the shaft and set up how it is supported and loaded.

3. Click the check box next to the Material callout. This will open a Material Types dialog box.

4. Select Steel, and click OK to make it the active material.

5. In the 2D Preview, click the Free Support icon at the far-left end of the preview, and drag it to the middle of the rightmost section. See Figure 9.9.

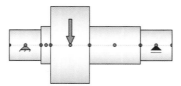

FIGURE 9.9 The 2D preview shows how the shaft is loaded and supported.

6. Click the Load icon, and the Loads & Supports area will change to show the type of load that is being used.

7. Double-click the Load icon to open the dialog box, showing the properties of the load.

8. Change the force to **200 N**, and click OK to close the dialog box.

9. Click the Calculate button, and then click the Graphs tab.
   The Shear Force graph shows the results.

10. Click the Deflection graph in the Graph Selection area to see that the deflection of this shaft is 0.811 micron (Figure 9.10).

11. Click OK to close the dialog box.

The Calculation tab varies based on the tool, but the information that you can enter or extract is precisely the information you would need in order to use engineering criteria to evaluate the components you're creating. You won't use the Calculation tools for all the Design Accelerators we will review, but keep in mind that they are available.

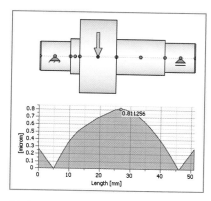

FIGURE 9.10 The Graphs tab displays the results of the calculations done on the shaft.

Using the Spur Gears Component Generator

The Design tab of the Spur Gears Component Generator dialog box is broken down into three major sections. The top is the Common area where you specify the tooth characteristics of the gear. The middle is where you set the requirements for size of the gears that support the teeth, and the bottom is an information area that will offer feedback on the design and calculations.

1. Make certain that the 2014 Essentials project file is active, and then open the c09-06.iam file from the Assemblies\Chapter 09 folder.

2. Hold the Ctrl key, and click the Spur Gear tool on the Power Transmission panel of the Design tab.
   The Spur Gears Component Generator dialog box opens and displays the default values (Figure 9.11).

Spur Gear

3. Beginning in the upper left, change Design Guide to Number Of Teeth.

4. Set the Desired Gear Ratio value to 4.0000 ul from the drop-down.

5. Set Module to .8 mm and Center Distance to **50.000 mm**.
   Now you need to make changes to the specifications for the individual gears.

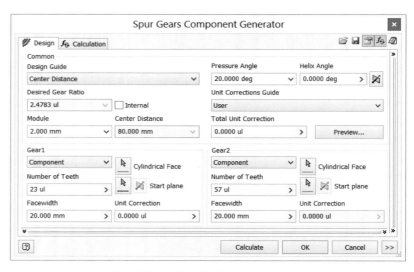

FIGURE 9.11 The default values for the Gear Generator

6. Choose Feature from both the Gear1 and Gear2 drop-downs. The Component option will build a new part in the assembly that can be fitted to the shaft. The Feature option will modify the shaft, and the No Model setting is used as a reference to develop the gear that mates to its values.

7. Set the Facewidth value for both gears to **13 mm**, and your dialog box should now look like Figure 9.12.

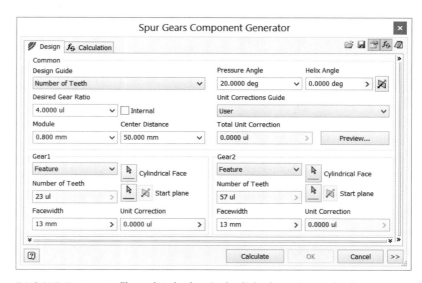

FIGURE 9.12 The updated values in the dialog box prior to selecting the geometry

8. Under Gear1, click the selection icon for the cylindrical face, and then select the cylindrical face on the largest-diameter segment of the gray shaft.

9. Click the selection icon for the start plane of Gear1, and click the flat face on the same shaft segment.

10. Start the Cylindrical Face selection for Gear2, and click the large, cylindrical face on the green shaft.

11. For the start plane of Gear2, click the large face on the same segment of the green gear.

12. Click the Calculate tool to update the geometry preview.
    The lower section will indicate a design failure. You might have to select the double arrows to expand the dialog box to see the error in the lower section.

13. Click OK to generate the gear features on the existing shafts, and click Accept to generate the gears (Figure 9.13) even with the incomplete information.

FIGURE 9.13 The updated shafts with the gear features added

Anyone who has created gears in 2D or 3D by developing a tooth profile and patterning it will appreciate the power of the Spur Gears Component Generator. Along with spur gears, there are design accelerators for worm and bevel gears.

Using the Key Connection Generator

Keys are used to align components, but you typically need material removed from at least two parts to be able to place the key. In this example, a gear component has been added to the assembly centered on the shaft.

1. Make certain that the 2014 Essentials project file is active, and then open the c09-07.iam file from the Assemblies\Chapter 09 folder.

2. Double-click the large gear.
   The gear will be solid with no opening for the shaft or accommodation for reducing the weight of the gear. Since the part is a solid model, you can use any solid modeling tool to change it, but in this example, you will create the shaft accommodation using the Key Connection Generator.

3. Click the Return icon on the Ribbon or click Finish Edit in the marking menu to return to the assembly.

4. Switch the Ribbon to the Design tab, hold the Ctrl key, and click the Key Connection tool in the Power Transmission panel.
   The Parallel Key Connection Generator dialog box (Figure 9.14) has some similarities to the Bearing Generator dialog box. The process is straightforward. Select the type of key you want, and then tell it what kind of groove you want for the shaft and hub or if you want one.

5. At the top of the dialog box, the current key standard is displayed. Click the down arrow at the end of this space to open the selection dialog box.

6. At the top of the dialog box showing the available key types, click the Standard drop-down, and select ANSI to limit the options.

7. Click the Rectangular or Square Parallel Keys type.

8. Just below the key selection is the value of the shaft diameter. Change this from 20.000 mm to **12 mm**.
   The next several steps define where the key is positioned and which parts of the assembly will be changed. Selecting the correct geometry is critical. Figure 9.15 shows where to click to properly locate the key.

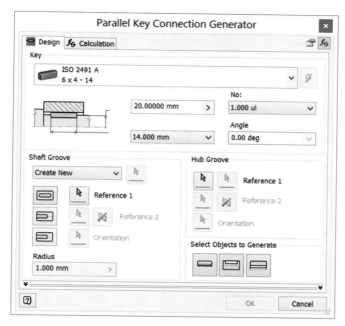

FIGURE 9.14 The Parallel Key Connection Generator dialog box default values

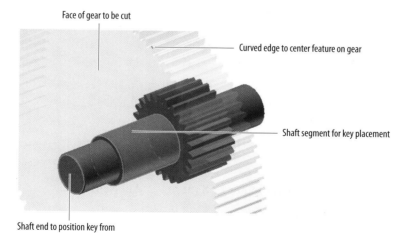

FIGURE 9.15 Select only the faces and edge (step 4) indicated.

In the dialog box, you can choose whether to generate the hub groove, shaft groove, or key.

9. Click the selection icon for Shaft Groove Reference 1, and then click the shaft segment between the gear and the smaller end segment.

10. The selection for Reference 2 automatically activates. Click the end of the shaft.

11. Reference 1 for the hub groove is now active. Click the face on the large gear.

12. For Reference 2 of the hub groove, you need to select a curved edge. Carefully select the tip of one of the gear teeth. If you click a side edge, simply click the selection icon and try again. You can zoom in to make it easier.

    Once the fourth selection is made, a preview of the key and grooves appears. You want to make one change to the shaft groove before you proceed.

13. In the Shaft Groove group, click the groove with one rounded end option.

14. There are two arrows on the ends of the preview of the key. Drag the left arrow roughly 11 mm from the origin to the right.

15. Drag the right arrow until the key changes to its 11.113 mm standard size, as shown in Figure 9.16.

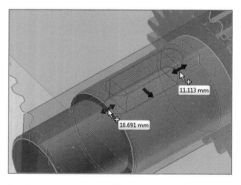

FIGURE 9.16 The position and length of the new key can be changed by dragging icons in the preview.

16. Click OK to generate the geometry, and click OK again to approve the creation of the new files. See Figure 9.17 for the finished key component.

FIGURE 9.17 The finished key with the groove built into the shaft

Both the shaft and the hub will have new features added to them, and a new part will be created for the key. Editing the key's size will update all of the related features.

Working with Additional Assembly Tools

Constraints and design accelerators are great, but they're not the only capabilities Inventor provides for an assembly. Some of these tools are not even unique to the assembly environment, but they simplify the process of defining your products.

For the next exercises, you will use a gear housing that is machined from a casting. As you start, you have a casting for only one half of the part, and it may need reinforcement.

Mirroring Components

When modeling parts, taking advantage of symmetry or the ability to pattern a feature can save work and improve quality. The same is true when it comes to whole components that are opposites of one another. For these components, you can mirror the entire part. In the context of an assembly, you can mirror several parts at once and generate the new files for them.

1. Make certain that the 2014 Essentials project file is active, and then open the c09-08.iam file from the Assemblies\Chapter 09 folder.

2. Start the Mirror Components tool from the Pattern panel of the Assemble tab.

3. The Mirror Components: Status dialog box will be prepared to select components to be mirrored. Click the housing and the two bearings in the Design window.

4. Click the selection icon for the mirror plane, and click the XY plane from the Origin folder of the assembly.

   A preview of the mirrored components will appear, and they will be color-coded based on the action to be taken. Green components will create a new file that is a mirror image of the original. Yellow components are new instances of existing geometry. This is the default state for standard components such as bearings.

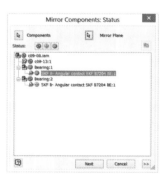

5. In the dialog box, click the yellow icon next to the first bearing.

6. The icon turns gray, which indicates that the components will be ignored by the tool.

7. Your screen should look like Figure 9.18. Click Next.

> The Ignore option in the Mirror Components dialog box can be used to skip parts accidentally clicked with the window or crossing selection.

FIGURE 9.18 A preview of the mirrored and instanced components

The Mirror Components: File Names dialog box allows you to specify the filename and path for the components that are to be created. You can also apply a new naming scheme to all the new components by adding a prefix or suffix to them.

8. Click OK to create the new files and produce the new components in the assembly, as shown in Figure 9.19.

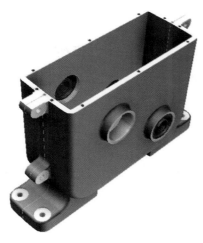

FIGURE 9.19 The mirrored housing in the assembly

Now that you have the mirrored part, you can make changes to the new part to make it uniquely different from the original, if need be. It is important to use the side that is least likely to need uniquely different features as the basis for the mirrored component to keep from having to remove geometry not needed in the copy.

Deriving Components

The Derive Component tool is one of the most powerful in Inventor. It can be used for everything, including linking the value of one parametric part into another, creating a mirror of a component, and, as in this exercise, using one part as the basis for another. The concept is to begin with the geometry, sketches, dimensions, or faces from one part and use them in another to ensure consistency.

1. Make certain that the 2014 Essentials project file is active. From the Quick Access toolbar, start the New File tool, and create a new file using the Standard (mm).ipt template.

2. Switch the Ribbon to the Manage tab, and start the Derive Component tool in the Insert panel.
   Derive is also on the 3D Model tab in the Create panel.

Derive

3. Use the Open dialog box to select c09-12.ipt from the Parts\Chapter 09 folder.

4. Click Open to go to the next step.

   The Derived Part dialog box (Figure 9.20) will allow you to choose what form your linked data will take. You can even choose to import only faces, sketches, or parameters. Once you've selected how you want to bring it in, you can change its scale or even mirror it around a plane.

FIGURE 9.20 The Derived Part dialog box offers complete control over the reuse of existing data.

5. Leave the default options, and click OK to completely link the c09–12 part into the new part.

   You can now add features to the linked geometry to make a machining of this cast part. Now, let's open an existing machining and the original casting.

6. Open the c09-13.ipt file from the Parts\Chapter 09 folder.

   Take a moment to review the holes, extrusions, and other features that have removed material from the original casting.

You can use Derive Component to create a single or multibody part out of an assembly even as a mirrored or scaled geometry.

7. Open c09-12.ipt from the Parts\Chapter 09 folder.

8. In the Browser, locate the End Of Part marker. Click and drag it to the bottom of the list of features.

   After the part updates, you will see that a rib feature has been added to the model.

9. Click the File tab below the Design window to display your new part. It will not show the change to the casting yet.

10. Click the local update icon in the Quick Access toolbar to bring the changes into your new part.

11. Switch your Design window to the c09-13.ipt file and update it. See Figure 9.21.

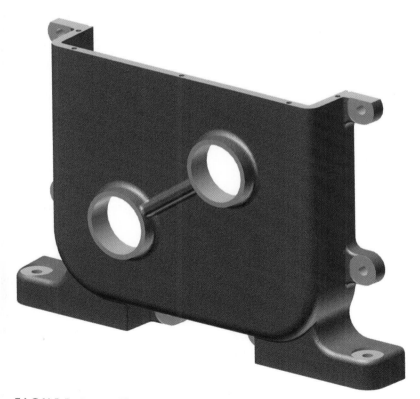

FIGURE 9.21 The updated part showing changes made to the casting file

T I P   If you use a single casting as the base for multiple components, the Derive Component tool will save you countless hours of updating machined parts. As mentioned in the introduction, this tool has many more uses and warrants further exploration.

Constraining and Animating Assembly Motion

Certification
Objective

In Chapter 5, I covered the majority of Assembly constraint options that hold components in place. The Motion constraint can bind the movement of one component to another.

1. Make certain that the 2014 Essentials project file is active, and then open the c09-09.iam file from the Assemblies\Chapter 09 folder.

2. In the Relationships panel of the Assemble tab, click the Constrain tool.

3. In the Place Constraint dialog box, click the Motion tab.

4. Set the Ratio value to 4, and then set the Solution value to Reverse. This tool is click-sensitive. The ratio is applied from the first selection to the second selection.

5. For the first selection, click a flat face or the end of the brass gear.

6. For the second selection, click the end of the gray shaft below it, as shown in Figure 9.22.

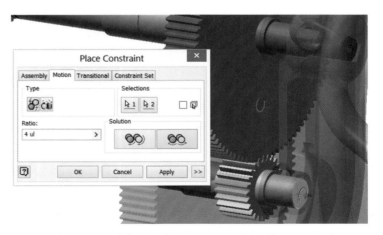

F I G U R E   9 . 2 2   Selecting the components that will move together

7. Click OK to create the constraint.

8. Click and drag either the wheel or the wooden handle to see the mechanism work. Note that the last shaft turns at a final ratio of 16:1 over the handle.

9. Locate the Drive This Constraint item in the Browser under the New Shaft part.

   Even though this constraint is suppressed, it can still be used to animate the assembly.

10. Right-click the constraint, and select the Drive tool from the context menu.

11. In the Drive Constraint dialog box, set the End value for the range of motion to **360**.

12. Expand the dialog box, and set the value for Repetitions to 3, as shown in Figure 9.23.

You can save the animation of a Driven constraint to an AVI or WMV file by clicking the Record button in the dialog box.

F I G U R E   9 . 2 3   Driving a constraint can show how a mechanism works.

13. Click the Forward button to see how the mechanism works.

Applying motion constraints to an assembly helps you verify that your design works the way you need it to work. Being able to animate this action can help you show others what you intend the design to do.

THE ESSENTIALS AND BEYOND

Representations can add control over how your assemblies look, but leveraging the LOD tools can improve the performance of your computer, save you time waiting for changes to update, and save you money by keeping your hardware effective longer.

Design accelerators bring a whole new dimension to CAD, adding engineering tools to common component tools. Integrating these calculations may help you improve the efficiency of your products.

ADDITIONAL EXERCISES

▶ Practice using LOD representations in your day-to-day work. These tools provide many advantages, including speeding up the drawing view by using them as the basis for view creation.

▶ Try using the Cam, Spring, and other tools available on the expanded Power Transmission panel.

▶ Explore different shaft loads and loading options to see how the shaft reacts.

▶ Look at the Calculation tab on all the design accelerators as you try them to see what information you will need to supply to allow the calculators to work.

CHAPTER 10

Creating Sculpted and Multibody Parts

In this chapter, a majority of the Autodesk® Inventor® 2014 tools you use appear in the Plastic Part panel. It's important that you explore these features for other uses ranging from vents for sheet metal parts to intelligent fillet placement on traditional solid models. Other techniques — such as building solids from surfaces and using multibody modeling — can be employed rather than constructing solid extrusions or building an assembly from separate components.

Think about complex models that you need to create and imagine using the tools from this chapter to approach them in a new way.

▶ **Developing specialized features for plastic components**

▶ **Creating an assembly using a multibody solid to maintain a consistent shape**

Developing Specialized Features for Plastic Components

Plastic part features are specialized tools that create geometry that you *could* create using traditional features. But if you went the traditional route, you'd need to use several features to create the same geometry that is common in these types of parts. Figure 10.1 shows the model you'll complete in this chapter's exercises. To create and refine its shapes, you'll use many of the most common sculpting tools available in Inventor.

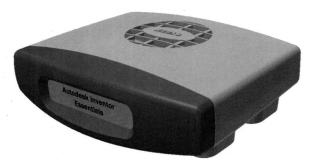

FIGURE 10.1 The completed model from this chapter

Creating a Surface Using a Sketched Feature

Not every solid model has to start out as one. In fact, when approaching a design with a number of complex surfaces you may find it much more efficient to build that model as separate, complex surfaces. Solid modeling tools that use sketches can be used to create surfaces as well as solids. In the following steps, you will add a surface that completes the envelope of a part that will be converted to a solid model and then detailed.

1. Verify that the 2014 Essentials project file is active, and then open the c10-01.ipt file from the Parts\Chapter 10 folder.

2. Choose the Revolve tool from the Create panel.
   Because this is an open sketch the mini-toolbar for the Revolve tool will reflect that Inventor assumes you are creating a surface feature rather than a solid.

3. Select the green construction line as the axis.
   Selecting the axis will generate a preview of a fully revolved surface.

4. Use the pull-down menu to select Angle.

5. Set the angle to 40 degrees and change the direction option to Symmetric.

6. When your preview looks like Figure 10.2, click the green check mark for OK or right-click in the graphics window and click OK in the marking menu to generate the surface.

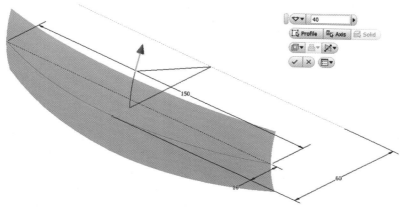

FIGURE 10.2 Surfaces can be created using common solid modeling tools.

7. In the Browser, click and drag the End of Part marker down the list of features to see the part's features.

All of the features of this part — the Loft, Boundary Patch, Extrusion, and even patterns — are all surfaces. In the next step you will combine these surfaces into a solid model.

Sculpting a Solid from Surfaces

In Chapter 7, "Advanced Part Modeling," you used the Replace Face tool to replace a contoured face. The Sculpt tool could be considered an evolution of Replace Face, but it is far more powerful. It will help you combine a series of overlapping surfaces into a solid model. In the following steps, you'll use it to create the initial shape of your model:

1. Verify that the 2014 Essentials project file is active, and then open the c10-02.ipt file from the Parts\Chapter 10 folder.

2. Click the Sculpt tool in the Surface panel of the 3D Model tab.

3. Select all the surfaces in the Design window, including the plane at the base of the part.

4. As the surfaces are selected, a preview of solid bodies appears. Once all the surfaces are selected, click OK.

The complete solid should look like Figure 10.3.

Selecting the direction arrows in the surfaces allows you to guide the direction of the volume that will be solidified if there is any ambiguity.

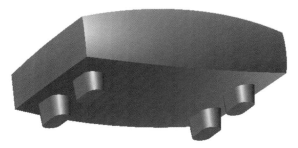

FIGURE 10.3   The solid formed by the surfaces

 TIP   When creating components with a series of complex surfaces, you may want to create the body as surfaces and sculpt them together rather than trying to construct a loft or other features.

Stitching Surfaces Together

You can combine surfaces to form a new surface, and then that surface can be used for other things. In this example, you'll use the Stitch tool to define a surface to be removed:

1. Verify that the 2014 Essentials project file is active, and then open c10-03.ipt from the Parts\Chapter 10 folder.

2. In the Browser, expand the Solid Bodies folder.

3. Right-click the icon next to Solid1, and deselect Visibility to turn off visibility of the solid model.

4. Click the Stitch tool in the Surface panel of the 3D Model tab.

5. Select the extruded surface and the boundary patch; then click Apply to create the new surface.

6. Click Done to close the Stitch dialog box.

7. Make the solid visible again.

8. Click the Sculpt tool, and select the new surface.

9. Switch to the Remove option.

A preview appears showing the portion to be removed in red (Figure 10.4).

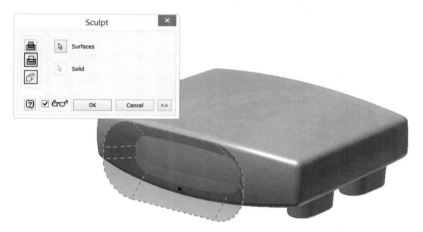

FIGURE 10.4 The Sculpt tool can also remove a portion of the solid using a surface.

10. Click OK to finish editing the part.

If you have a series of surfaces that meet at their edges, you can stitch them together into a solid. It will work, but cleaning up the edges of the surfaces is more difficult than allowing them to overlap and using the Sculpt tool.

 TIP You can turn a single part into separate bodies representing and even defining the multiple parts of an assembly. By forming these various bodies using the same geometry, you can maintain the consistency of their shapes.

Splitting Bodies

You can generate most sketched features as a new body by simply selecting the New Solid option in the dialog box used to create that feature (such as Extrude, Revolve, or Loft). You can also use a plane or sketch geometry to split a face, remove a portion of a body, or even divide a solid model into separated bodies. To use the Split tool in this model, you will use a plane that will define the boundary between the two bodies:

1. Verify that the 2014 Essentials project file is active, and then open c10-04.ipt from the Parts\Chapter 10 folder.

2. Select the Split tool in the Modify panel of the 3D Model tab.

3. Select the Split Solid option in the Split dialog box, and click the visible work plane on the part or the Splitting Plane 1 feature in the Browser.

4. Click OK to break the part into two bodies.

5. In the Browser, expand the Solid Bodies folder to see the new solids that have been added.

6. Rename Solid2 to **Front** and Solid3 to **Back**.

7. Click the solids individually in the Browser, and change their colors using the drop-down menu in the Quick Access toolbar. My choices are shown applied to the solid in Figure 10.5.

FIGURE 10.5 Once the solid is split into multiple bodies, you can treat them differently.

You can divide and subdivide a body as many times as needed to suit your needs. You can also use an extrusion to cut a part and then use the shared sketch to create a new body to form interlocking components.

Adding a Lip

A lip allows the plastic parts to remain aligned so the form remains consistent. You can use other sketched features, such as the Sweep tool, to make the geometry, but the Lip tool makes it easier to adjust things like the draft options and to recess edges.

1. Verify that the 2014 Essentials project file is active, and then open c10-05.ipt from the Parts\Chapter 10 folder.

2. Rotate the front body so that you can see the back side of it.

3. Select the Lip tool in the Plastic Part panel on the 3D Model tab.

4. In the Lip dialog box, make sure the option on the left is set to Lip, and then select the Pull Direction check box.
   This changes the dialog box options for constructing the lip feature.

5. Click the flat face around the perimeter of the part to set the Pull Direction option.

6. Once the direction is selected, click the inner edge of the perimeter for the path edges, as shown in Figure 10.6.

FIGURE 10.6 Selecting model edges to place a lip feature

7. Click the Lip tab in the dialog box, and review the options.

8. Click OK to place the feature using the default options. The result is shown in Figure 10.7.

FIGURE 10.7 The lip added to the front body

T I P   You can also select the back part and place a lip using the Groove option. If a lip feature is placed in a part, its properties will be picked up by a groove in the same part.

Adding a Boss

Certification Objective

Like the Lip tool, the Boss tool comes with opposite geometry sets. The head and thread sides of the boss are collections of complex geometry that would normally take several features to create.

The bosses you add in this exercise will allow you to assemble the halves of the back together using screws:

1. Verify that the 2014 Essentials project file is active, and then open the c10-06.ipt file from the Parts\Chapter 10 folder.

2. Select the Boss tool in the Plastic Part panel on the 3D Model tab.

3. Make sure the Head option is active. It is the button in the upper-left corner of the screen.

4. Switch the Boss dialog box to the Head tab, and change the bottom two size options from 6.6 to 6 and from 7.54 to 7.

5. Expand the Draft Options section in the dialog box.

6. Change the first two draft options from 2.5 to 2 degrees. Refer to Figure 10.8 for the dialog box values.

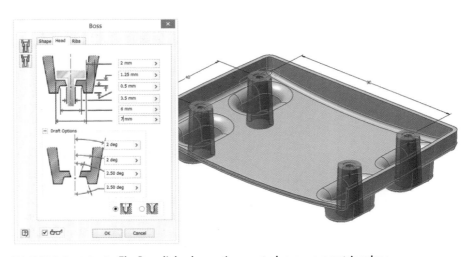

F I G U R E   1 0 . 8   The Boss dialog box options control many parametric values.

7. Click OK to generate the Head side bosses.

8. Expand the `Solid Bodies` folder, right-click the Back–Top body, and select Hide Others in the context menu.

     This makes a new body visible and turns off the others. Now, you can add features to this body.

You can also build ribs around the boss, and you can adjust their height, thickness, and angle.

9. Scroll down in the Browser, and expand the Boss1 features.

10. Click and drag Sketch9 just above Boss1 so that it is now shared.

11. Right-click the sketch, and make it visible so that it can be used to create a new feature.

12. Click the Boss tool, and switch it to the Thread option.

13. Set the Fillet value to 3 (Figure 10.9), and click OK to place the new features.

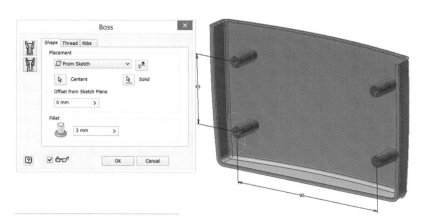

FIGURE 10.9 A boss can automatically add a fillet at the base for the new boss.

 TIP You can add stiffening ribs as part of the feature. You can also allow the hole to pass all the way through the component or be created only to a specified depth. By default the bottom of the hole will be set to and follow the contour of the existing face to reduce the amount of sink on the part.

Creating a Rest

A *rest* is like an extrude command that understands a wall thickness. It is meant to build a flat spot and can construct the internal and external faces at the same time. Now you will use the Rest tool to add a pocket to the bottom of the housing.

1. Verify that the 2014 Essentials project file is active, and then open c10-07.ipt from the Parts\Chapter 10 folder.

2. Select the Rest tool in the Plastic Part panel of the 3D Model tab.

3. After the preview appears, click the direction option in the dialog box on the left to put the hollow of the rest to the inside of the part.

4. Set the thickness to 2, and then click the More tab.

5. Set the Landing Taper and Clearance Taper values to 2, as shown in Figure 10.10, and click OK to create the rest.

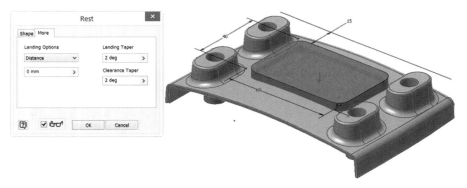

F I G U R E 1 0 . 1 0   The Rest tool will create a flat on the plastic part.

T I P   You can also place a rest that intersects a face; in that case, it would simultaneously add and remove geometry in different directions.

The Rule Fillet Tool

The Fillet tool used in Chapter 3, "Introducing Part Modeling," and Chapter 7 is powerful but can require a lot of input from the user. The Rule Fillet tool works based on relationships between the body of the part, its features, and its faces. In the following steps, you will use this tool to add fillets between a rest feature and the other features of a solid by specifying how the features blend, not the edges on which you want a fillet placed:

1. Verify that the 2014 Essentials project file is active, and then open c10-08.ipt from the Parts\Chapter 10 folder.

2. Select the Rule Fillet tool in the Plastic Part panel of the 3D Model tab.

   The primary options for the tool allow you to select either a feature or a face.

3. When the Rule Fillet dialog box opens, click the rest feature, set Radius to 2, and leave the rule set to Against Part.

   The fillet preview appears anywhere the selected feature makes contact with the rest of the part.

4. Click below the first rule where it says *Click to add* to create a new rule.

5. Click the rest again; leave the radius at 1 mm but change the rule to Free Edges. See Figure 10.11.

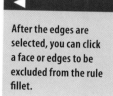

After the edges are selected, you can click a face or edges to be excluded from the rule fillet.

FIGURE 10.11 Adding a rule fillet between a feature and the part

   This applies the second rule to any edge that is not tangent with another face or already slated to be affected by another rule.

6. Click OK to place the fillets based on these rules.

TIP A rule fillet placed between features or between a feature and the body does not cause an error message if the conditions don't allow it to generate. Instead, it lies dormant until the conditions are restored and it can generate the feature again.

Adding a Grill

Expanding the Grill dialog box reveals a Flow Area calculation. This refers to the opening created by the grill, because grills are commonly used for ventilation.

A *grill* is a feature that creates an opening in a part. It frequently has ribs across the opening. It can also include areas of solid obstruction called *islands*.

1. Verify that the 2014 Essentials project file is active, and then open c10-09.ipt from the Parts\Chapter 10 folder.

2. In the Design window, orbit the part so you can clearly see the sketch that is visible.

3. Select the Grill tool in the Plastic Part panel of the 3D Model tab.
   The Grill dialog box has five tabs with options for the sketch geometry you want to use for that subfeature.

4. The first tab sets the boundary. To define the boundary, click the red circle in the sketch.

5. Click the Island tab, and for its profile, select the green ellipse in the middle of the sketch. Set the thickness of the island to 1.

6. Select the Rib tab, and then click the six blue parallel lines in the sketch. Do not change any of the dialog box options.

7. Click the Spar tab, and pick the three yellow lines that are perpendicular to the last lines. Set Spar Thickness to 2 and Top Offset to .5.

8. Compare your model to Figure 10.12.

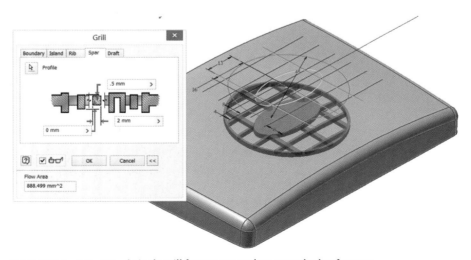

FIGURE 10.12 A single grill feature can replace several other features.

9. Expand the dialog box, and it will update to show what the open sectional area of the grill is, for ventilation considerations.

10. Click OK to generate the grill.

The grill is the most flexible feature for plastic parts. A wide selection of sub-features allows you to pick and choose the elements you want to place.

Embossing or Engraving

The rules for creating an embossed or engraved feature, such as the product logo, are the same as if you were going to extrude a solid feature. You can use a closed sketch or text to define the feature's geometry.

1. Verify that the 2014 Essentials project file is active, and then open c10-10.ipt from the Parts\Chapter 10 folder.

2. Position the model so you can see the text in the sketch above the grill feature.

3. Select the Emboss tool in the Create panel of the 3D Model tab.

4. When the Emboss dialog box opens, click the text.
    The Emboss tool has three options. One will add geometry, one will remove, and the third will do both if the sketch plane intersects a face on the part.

5. Keep the Emboss From Face option active, and set the depth to .5.

6. Make sure the direction indicator is pointed toward the part (Figure 10.13). It might be necessary to change it. Then, click OK to create the feature.

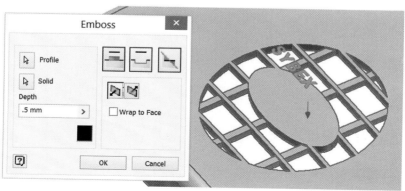

FIGURE 10.13 Adding text to the face of a part

T I P   The Emboss tool also has the ability to wrap a profile or text around a face so that the edges of the feature are normal to the surface they're wrapped on.

The Snap Fit Tool

The two options for the geometry for this tool are the hook and the loop. The features are created with a wizard, so all that is needed is a work point or a sketch point.

1. Verify that the 2014 Essentials project file is active, and then open c10-11.ipt from the Parts\Chapter 10 folder.

2. Orbit the model so that you can clearly see the open portion.

3. Select the Snap Fit tool in the Plastic Part panel on the 3D Model tab. A preview of two features appears, because there are two points in the sketch.

4. In the Shape tab of the Snap Fit dialog box, click the Flip Beam Direction button, and notice the change to the preview. If the original preview showed the hooks correctly positioned (protruding from the cover), click the Flip Beam Direction button again.

5. Switch to the Hook tab. Notice the green arrows; the arrows indicate the direction of the hook. One of them points inward. Click that arrow until both hooks are pointing outward.

6. Switch to the Beam tab of the Snap Fit dialog box.

7. Match your dialog box values to those in Figure 10.14, and click OK to build the features.

F I G U R E   1 0 . 1 4   The Snap Fit tool creates complex geometry easily.

The loop side of a snap fit is placed in the same way. In general, all the plastic features follow a common process that makes them easy to learn.

 TIP Another advantage of using a multibody solid is not having to reenter size information to create mating plastic features. The feature remembers the options changed in the active file.

Adding Ribs

The Rib tool is not a specialized plastic part feature, but it is common in plastic parts. Ribs are used in many types of parts, and the Rib tool can be used to create a rib or a web; you have the ability to control the width and depth. For this exercise you will build Ribs from the lines that define their center.

1. Verify that the 2014 Essentials project file is active, and then open c10-12.ipt from the Parts\Chapter 10 folder.

2. Orbit the model so that you can clearly see the open portion.
   Sketched in the interior of the front part is a series of lines that will locate the ribs. Those lines do not extend to the edges of the part and do not need to do so. The rib feature is capable of extending itself to fill in the geometry.

3. Select the Rib tool in the Create panel of the 3D Model tab.

4. Click the four lines that are partially obscured by the block in the middle of the part.

5. Set the thickness to 2. See Figure 10.15 for the preview of the feature.

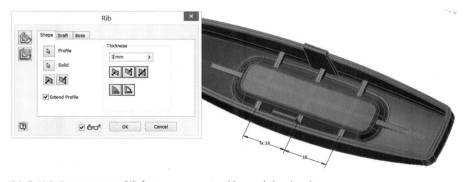

FIGURE 10.15 Rib features can extend beyond the sketch.

6. Click OK to create the ribs.

7. Orbit the part to see that the ribs did not penetrate through the front face of the part.

T I P   On parts with draft, it is difficult to properly create a sketch that doesn't risk creating a gap as it is created. Being able to create a rib with a sketch that doesn't meet the faces helps avoid problems when these parts change.

Adding Decals

It is possible to include geometry, text, or images in a sketch. A *decal* is a feature made from an image that is applied to flat or curved surfaces. Now you will use this tool to add detail to your design.

1. Verify that the 2014 Essentials project file is active, and then open c10-13.ipt from the Parts\Chapter 10 folder.

2. Create a new sketch on the XY plane.

3. Click the Image option in the Insert panel of the Sketch tab.

4. In the Open dialog box, navigate to the Parts\Chapter 10 folder, click the c10-15.png file, and then click Open.

5. Click a place in the Design window. Press Esc to finish placing.
   The size of an image can be controlled by a parametric dimension as well.

6. Place a dimension on the top of the image file, and set the value to 60. Press the Esc key to finish the dimension tool.

7. Use Zoom All (if necessary) to find the image in the sketch.

8. Drag the sketch so it is roughly centered on the recessed face of the part.
   You can also use dimensions and constraints to precisely locate decals by modifying the sketch the image is a part of.

9. Finish editing the sketch.

10. On the 3D Model tab, expand the Create panel by clicking the arrow next to the panel name, and select the Decal tool.

11. Uncheck the Chain Faces option in the Decal dialog box, and then click the image and the flat face behind it. See Figure 10.16.

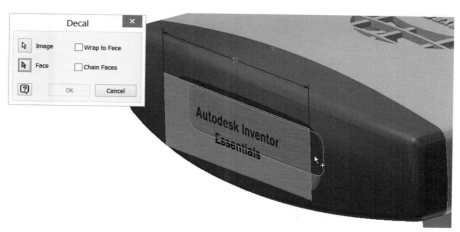

FIGURE 10.16   Selecting the face onto which to project the decal

12. Click OK to place the feature.

 TIP   This was a simple placement of a decal. The tool doesn't shine until you start wrapping images around faces. The Decal tool is capable of that as well.

Creating an Assembly Using a Multibody Solid to Maintain a Consistent Shape

A multibody part simulates an assembly. It can be used to create an assembly. The assembly will be made up of components created from some or all of the bodies in the part. Any change to the bodies of the part will be reflected in the part created from it.

Converting Bodies to Components

Converting the bodies to parts requires the user to specify a new assembly, select the bodies, and name the new part files:

1. Verify that the 2014 Essentials project file is active, and then open c10-14.ipt from the Parts\Chapter 10 folder.

2. Switch the Ribbon to the Manage tab, and select the Make Components tool from the Layout panel.

3. When the Make Components dialog box opens, it is in selection mode. Click the three components from the Design window or from the Solid Bodies folder of the Browser.

4. In the dialog box, set the Target Assembly Name field to **c10-14 .iam** and the Target Assembly Location field to **C:\Inventor 2014 Essentials\Assemblies\Chapter 10**.

5. Compare your dialog box to Figure 10.17, and then click Next.

FIGURE 10.17 Set the name of the assembly and its path.

6. You don't need to change anything in the Make Components: Bodies dialog box. Click OK.

You have now created the new files and opened the new assembly with the related parts. The parts will be grounded into position, so applying assembly constraints isn't needed.

Draft Analysis

There are tools for several types of shape and geometry analysis in Inventor. You can analyze the curvature, the sections of the body, and even the continuity of the surfaces.

Draft analysis gives a basic understanding of how a part might separate from a mold. It gives you the opportunity to make changes early in the process. In these steps you will use this tool to analyze the part.

1. Verify that the 2014 Essentials project file is active, and then open c10-15.ipt from the Parts\Chapter 10 folder.

2. Switch the Ribbon to the Inspect tab, and click the Draft tool in the Analysis panel.

Draft

3. Expand the Origin folder in the Browser, and select the XY plane for the direction to which the pull would be normal.

   In the Draft Analysis dialog box, you will find a scale showing the colors related to the draft angle. You can change the range of the positive and negative draft results and establish the range of allowed angles. Understanding the color related to the angle will help you interpret the results.

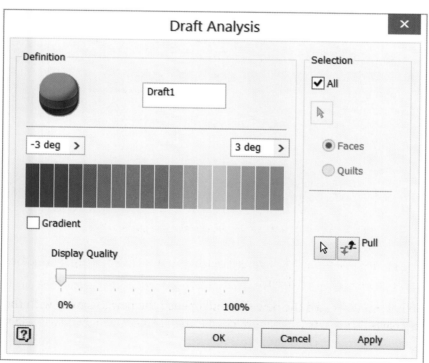

4. Click OK to close the Draft Analysis dialog box and display the results of the analysis, as shown in Figure 10.18.

FIGURE 10.18 Use draft analysis to see whether your part can be released from an injection mold.

5. After inspecting the results on the part, find the Analysis folder in the Browser, right-click, and uncheck Analysis Visibility in the context menu.

TIP This tool does not analyze the shrinkage of material or do any material flow analysis. Those tools are available in more advanced versions of Inventor.

THE ESSENTIALS AND BEYOND

Working with the specialized plastic part features, you learned how a common feature can have its options and variables built into a simple dialog box.

Multibody parts have many applications beyond plastics. They can be used any time more than one part needs to share a common shape.

ADDITIONAL EXERCISES

▶ Apply the groove side of the lip to the back body in c10-05.ipt.

▶ When you are going to apply edge fillets between features, try using rule fillets instead.

▶ Use the Rib tool to create ribbing at an angle or across openings by defining a depth.

▶ The Emboss tool can also cut material. Try doing the exercise for the Emboss tool using a cutting option instead.

CHAPTER 11

Working with Sheet Metal Parts

Sheet metal fabrication follows rules. For bending operations, certain things are not allowed. For example, you can't fold material if you cannot cut the part in the flat pattern, and material comes in certain standard thicknesses. Therefore, you can greatly simplify the design of sheet metal components by including fabrication considerations when working with the design tools.

It's entirely possible to construct sheet metal parts using solid modeling tools, such as Extrude and Revolve. However, working with specialized tools eliminates steps, and because these tools work with styles, changing the style that the part is based on updates the part.

In this chapter, you will create different parts using the sheet metal tools of the Autodesk® Inventor® 2014 software. The operation of these tools is similar to other solid modeling tools, so they should be easy to learn.

▶ **Defining sheet metal material styles**

▶ **Building sheet metal components**

▶ **Preparing the part for manufacture**

▶ **Documenting sheet metal parts**

Defining Sheet Metal Material Styles

Sheet metal defaults are styles used to establish consistency between components. You can create these styles within a part (with a scope limited to that part) or save them into a template to be available as you create new parts. A part created using one of these styles will update its physical attributes if the active style of the components is changed. In the following steps, you will:

1. Be sure that the 2014 Essentials project file is active, and then open the file c11-01.ipt from the Parts/Chapter 11 folder.

2. Click the Sheet Metal Defaults tool (Figure 11.1) in the Setup panel of the Sheet Metal tab, and click the edit icon next to the Sheet Metal Rule drop-down.

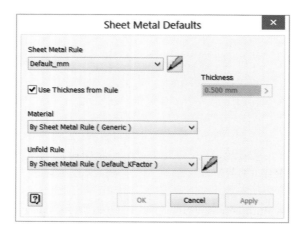

FIGURE 11.1 The Sheet Metal Defaults dialog box specifies whether you're basing your part on a predefined rule or what settings to use if you're not.

3. When the Style And Standard Editor dialog box opens, make sure the Default_mm style is active, and click the New button at the top.

4. In the New Local Style dialog box, enter the name Aluminum 3 mm, and click OK.

5. In the Style And Standard Editor dialog box, change the Material drop-down to Aluminum 6061.

6. Set Thickness to 3 mm and the Miter/Rip/Seam Gap value to Thickness / 2.

7. Set Flat Pattern Punch Representation to Center Mark Only. See Figure 11.2.

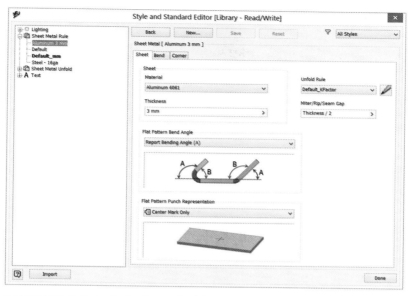

FIGURE 11.2 On the Sheet tab of the Style And Standard Editor, you define the thickness and material of the sheet.

8. Click the Bend tab, and change Relief Shape to Tear and Bend Radius to 2 mm, as shown in Figure 11.3.

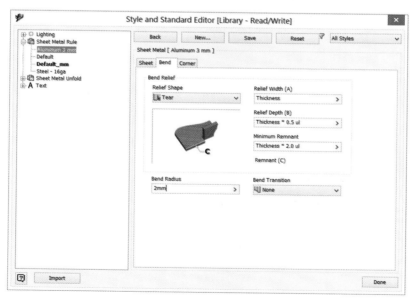

FIGURE 11.3 Use the Bend tab to control the radius of the bend and how edges will intersect with the sheet.

9. Click the Corner tab. Under 2 Bend Intersection, change Relief Shape to Linear Weld.

10. Change the Relief Shape value for 3 Bend Intersection to Full Round. See Figure 11.4.

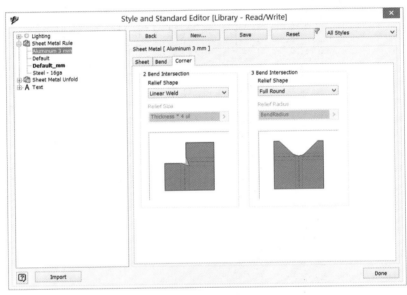

FIGURE 11.4 The Corner tab regulates how material will be removed from the corner or whether it will be left in.

11. Click Save, and then double-click the Aluminum 3 mm rule (on the left) to make it active.

By creating a series of sheet metal defaults and saving them in a template, you can save time by not re-creating the styles over and over. Later in this chapter you will update the completed part to a new specification using a different style.

Building Sheet Metal Components

Even though the Inventor sheet metal tools are limited to components that can be made with break press operations, the variety of components that can be made with this process is enormous. As a result, there are a large number of sheet metal tools to accommodate the various features that are commonly needed.

Creating a Basic Face

The Face tool looks much like the Extrude tool, but with sheet metal features, there is no need to input a thickness from the material because that is determined by the sheet metal default style. Now you will see just how easy the incorporation of styles makes creating a Face feature.

Certification
Objective

1. Verify that the 2014 Essentials project file is active, and then open c11-02.ipt from the Parts/Chapter 11 folder.

2. Select the Face tool from the Create panel of the Sheet Metal tab.

3. There is more than one closed profile, so no profile is automatically selected. Select the larger portion of the sketch, as shown in Figure 11.5.

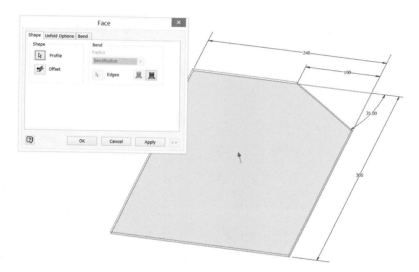

FIGURE 11.5 Select the portion of the sketch from which to create the face.

4. Click OK to create the face.

TIP   A face feature is frequently the base feature for sheet metal parts. You should use it to define the overall shape of your part. Face features can be added to a part at any time.

Adding Sides to the Part

Certification
Objective

Flanges are typically features that are bent from the base. The Flange tool includes nearly every conceivable option for creating your features. Follow these steps to create multiple Flanges in just a few steps:

1. Verify that the 2014 Essentials project file is active, and then open c11-03.ipt from the Parts/Chapter 11 folder.

2. Select the Flange tool in the Create panel of the Sheet Metal tab or from the marking menu.

3. On the Shape tab, click the Loop Select mode, and click the front face of the part to highlight all edges and get a preview of the feature.

4. Set the Height Extents value to 40. If the flange preview is not in the same direction as Figure 11.6, hold the Shift key, deselect the lower edge, and then reselect the edge of the top face.

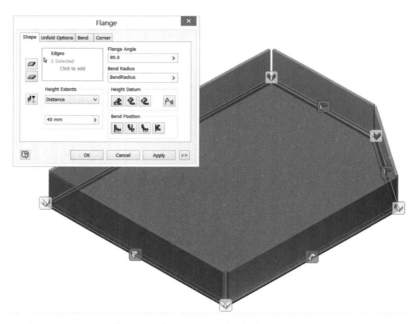

FIGURE 11.6 Loop speeds creating multiple flanges

TIP   When selecting edges for placing flanges, be aware of which edge you're selecting. The height is based on the edge you select. If you want the height measured from an outside face, you can reverse the direction of the flange to pass back through an existing face.

5. Click Apply to create the new flanges.

6. Switch to Edge Select mode, and set Height Extents to 20.

7. Click the inside edges of the previously placed flanges, as shown in Figure 11.7. If you cannot select an edge, click next to 0 Selected to enable selection mode.

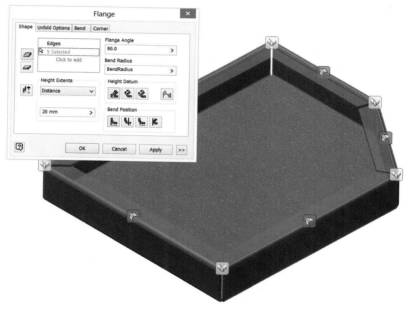

FIGURE 11.7 Corners that would otherwise overlap are automatically mitered.

8. Select the Corner Edit tool at the top-left corner of the part.

9. When the Corner Edit dialog box opens, click the Corner Override check box, and select Reverse Overlap from the drop-down. See Figure 11.8.

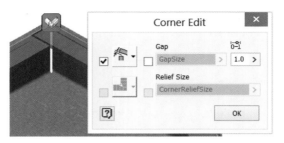

FIGURE 11.8 The automatic corner treatments can be overridden.

10. Click OK to accept the corner edit and then click OK to place the flanges.

11. Select the Flange tool again, and click the front edge of the new flange on the angled face. (See Figure 11.9.)

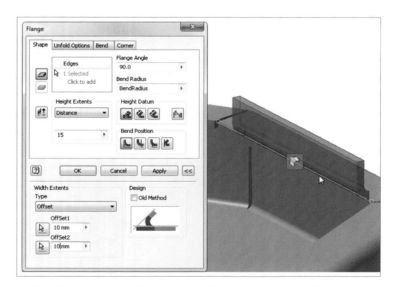

FIGURE 11.9 Adding a flange with the ends offset from the edge length

12. Expand the dialog box to see additional options. Under Width Extents, set Type to Offset, and change the offset values to 10 mm.

13. Change Flange Height to 15 mm, as shown in Figure 11.9, and click OK to place the feature.

The angle of a new flange and how it is developed in the footprint of the existing parts are all important options worth experimenting with when using the Flange tool.

Building from the Middle

At times, you might need to build faces in space or from projections from other parts in the assembly. Faces can be created in space with no connection to the rest of the part, but eventually they will need to be attached to the rest of the part. The Bend tool can help you do that using more than one workflow. In the following steps, you will use the face tool to build flanges as a short cut to stacking flanges on each other.

Certification
Objective

1. Verify that the 2014 Essentials project file is active, and then open c11-04.ipt from the Parts/Chapter 11 folder.

2. On the Sheet Metal tab, select the Bend tool from the Create panel.

Bend

3. Click one of the long edges of the red Face feature and then an edge on the nearest flange of the main body.
   The preview shows a feature that bridges the selected edges and cuts into the flange on the main part.

4. Click the Full Radius option in the Bend dialog box. Notice that the preview shows a feature that is tangent to both faces.

5. Return to the 90 Degree option, and click the Flip Fixed Edge option. Your screen should look like Figure 11.10.

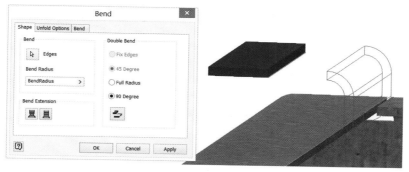

FIGURE 11.10 The Bend tool builds geometry between features.

This bases the geometry on the selected edge of the main part.

6. Click OK to create the new feature.

End of Folded

TIP   A face can also be used to add geometry to a part when the shape is too complex for a flange feature.

7. Move the End of Folded marker to the bottom of the Browser to expose a sketch.

8. Select the Face tool from the Create panel.

9. When the face preview appears, click the Edges selection icon, and click the same edge on the main body as you did in step 3.

10. After the dialog box expands, click the 90 Degree Double bend, and click the Flip Fixed Edge option.

11. Click OK to generate the feature; Figure 11.11 shows the result.

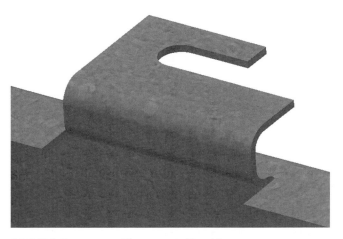

FIGURE 11.11   The generated bend feature

You can create bend features from an existing feature or while creating a new one. The other bend options, such as limiting the bend to 45 degrees or using Flip Fixed Edge, can build geometry that could take several features to make without them.

Updating a Part Using Styles

Now that the part is complete, you can change it to another material or material thickness without worrying about changing the position of features.

If you will be making the same component out of several materials, you should make sure that the material is added to the part away from the position of the critical dimensions. Follow these steps see how changing the style changes the part.

1. Verify that the 2014 Essentials project file is active, and then open c11-05.ipt from the Parts/Chapter 11 folder.

2. Click the Sheet Metal Defaults tool in the Setup panel of the Sheet Metal tab.

Sheet Metal Defaults

3. When the Sheet Metal Defaults dialog box opens, use the Sheet Metal Rule pull-down menu to select the Aluminum 3 mm style, as shown in Figure 11.12.

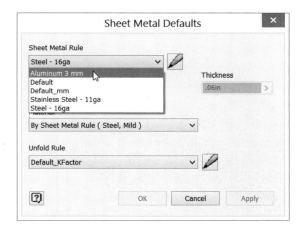

FIGURE 11.12 Changing the material of the part

4. Click Apply to update the part and see the result in the graphics window.

5. Switch the active Sheet Metal Rule to Stainless Steel - 11ga and click Apply.

   Not only will the part update its material thickness and appearance, the corner reliefs will change shape as well as the bend radius is applied to the edges.

6. Click the edit icon to the right of the Sheet Metal Rule to open the Style And Standard Editor dialog box.

7. In the column on the left, expand the Sheet Metal Unfold rules and click the Bend Table (in) option to see the table shown in Figure 11.13.

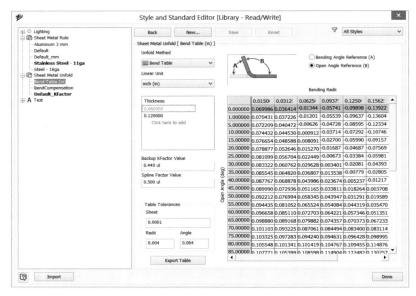

FIGURE 11.13 A bend table can be used instead of a K Factor.

The style definition for the Stainless Steel sheet uses a bend table rather than a simple K Factor calculation. This allows manufacturers to leverage the knowledge of their process to generate more precise flat patterns. The data for this table was imported from the stock data included in the Inventor installation under the `Public Documents` folder.

8. Click Done to close the dialog box.

If you work with sheet metal parts on a regular basis, it is possible that the fabricators you work with already have a bend table in one form or another that you can leverage to generate your styles specific to your tooling.

Using an Open Profile

To create a long part with a consistent profile, you could build a series of flanges attached to one another, but using the Contour Flange tool is a much simpler way to go, as you'll see in these steps:

1. Verify that the 2014 Essentials project file is active, and then open `c11-06.ipt` from the `Parts/Chapter 11` folder.

2. Click the Contour Flange tool in the Create panel on the Sheet Metal tab.

3. Select the profile, and the preview appears. Expand the Contour Flange dialog box if necessary and set the Distance value to 300.

4. Change the direction for the feature as shown in Figure 11.14.

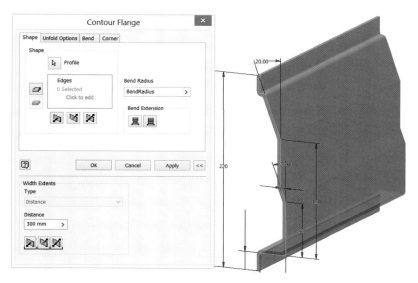

FIGURE 11.14 The contour flange adds geometry from an open profile.

5. Click OK to create the feature.

Without this tool, creating the part you've just built would have taken a face and seven flanges. Contour Flange adds the band radii at the corners of the sketch automatically. You can build curves into the sketch, and they will be maintained in the feature.

Adding Library Features around Bends

A challenging problem is dealing with features that transition across bends. Another challenge is working with features that you will use repeatedly. Punch tools are a common need in sheet metal, and you can define them and save them to a library for reuse. In these steps, you will add features across bends and have them follow the part's shape.

1. Verify that the 2014 Essentials project file is active, and then open c11-07.ipt from the Parts/Chapter 11 folder.

2. Select the Unfold tool from the Modify panel of the Sheet Metal tab.

3. When the Unfold dialog box opens, click the yellow face of the part for the stationary reference.

4. Once the stationary reference is selected, the bends of the part will be highlighted. Click the bend shown in Figure 11.15, and then click OK.

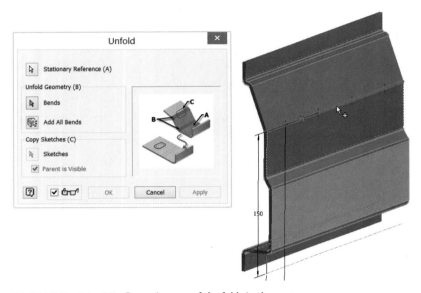

FIGURE 11.15 Removing one of the folds in the part

An Inventor punch tool feature can be placed only on a flat surface. By unfolding a portion of the part, you've created a flat surface large enough on which to place the feature.

5. Click the Punch Tool icon in the Modify panel.

6. In the Punch Tool Directory dialog box, find the Square Emboss.ide feature, and double-click it to start placing it on the part.

   Because there was a pattern of hole centers in the sketch, four previews appear for the Punch tool.

7. Switch to the Size tab of the PunchTool dialog box, and set the Length value to 60 mm and the Height value to 4 mm, as shown in Figure 11.16. (The preview will not update.)

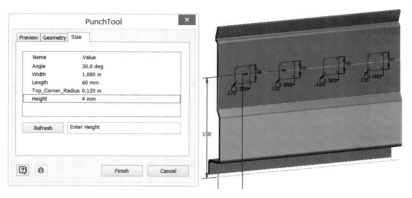

FIGURE 11.16 Placing the punch tools on the part

8. Click Finish to build the features.

   Now, you need to restore the part to its original shape. This requires the punch tools to be bent with the other geometry.

9. Click the Refold tool in the Modify panel.

10. Once again, use the yellow face as the stationary reference, and click the only unfold feature that highlights.

11. Click OK to restore the shape of the part, as shown in Figure 11.17.

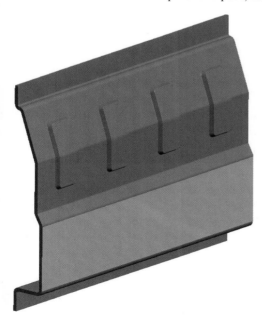

FIGURE 11.17 The part's bend restored

You can use Unfold and Refold on as many edges as you like if you need to measure or locate features off a different face. A similar technique can be used by projecting a preview of the flat pattern into the sketch, as you will do in the next exercise.

Adding Cut Feature around Bends

Creating a cut across one or more bends can be challenging, especially when you need to control the position of the ends of the cut on more than one face. Follow these steps to create a cut across bends without using the Unfold and Refold tools.

1. Verify that the 2014 Essentials project file is active, and then open c11-08.ipt from the Parts/Chapter 11 folder.

2. Edit the cut sketch by double-clicking it in the Browser.

3. On the Format panel, click the Construction override so new entities will be construction lines.

Project
Flat Pattern

4. Expand the Project Geometry tool in the Draw panel of the Sketch tab, and click Project Flat Pattern.

5. Click the face that the rectangle of the sketch overhangs.

6. Add a 15 mm dimension between the end of the rectangle and the end of the projection of the space between the punch tools, as shown in Figure 11.18.

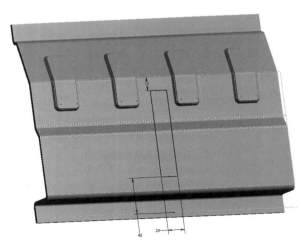

FIGURE 11.18 Projecting the flat pattern makes it possible to dimension a position after it will be bent.

7. Finish the sketch.

8. From the marking menu or the Modify panel of the Sheet Metal tab, select the Cut tool.

9. In the Cut dialog box, select the Cut Across Bend check box, and click OK to see the finished part in Figure 11.19.

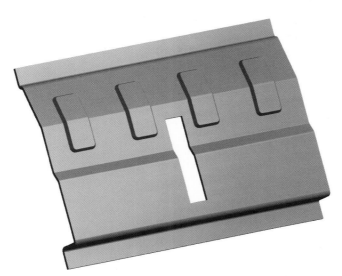

FIGURE 11.19 The position of the cut can be controlled on multiple faces.

Measuring the distance between the new cut and the edge of the face on which it ends shows that the distance includes the transition across the two bends.

Exploring an Advanced Open Profile Tool

The Contour Flange tool uses an open profile to generate a complex shape normal to the sketch. You can use the Contour Roll tool to revolve a profile around an axis to develop a sheet metal part.

1. Verify that the 2014 Essentials project file is active, and then open c11-09.ipt from the Parts/Chapter 11 folder.

2. Select the Contour Roll tool from the Create panel of the Sheet Metal tab.

The Contour Roll tool detects the single sketch and selects it, but you still need to select an axis.

3. Select the centerline to which the dimension of 200 is attached.

4. When the preview appears, change Rolled Angle to 60 deg and change the direction as shown in Figure 11.20. Click OK to generate the feature.

FIGURE 11.20 Bending the open profile around an axis

The Contour Roll tool is interesting, and it can create components that would be difficult to calculate using other tools.

Building Transitions in Sheet Metal

There are many uses for sheet metal components that change from one shape to another. Defining these shapes for manufacturing has been a longtime challenge for designers. In this exercise you will use the Lofted Flange tool to solve this problem.

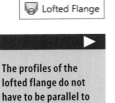

The profiles of the lofted flange do not have to be parallel to one another to create the feature.

1. Verify that the 2014 Essentials project file is active, and then open c11-10.ipt from the Parts/Chapter 11 folder.

2. On the Sheet Metal tab, click the Lofted Flange tool in the Create panel.

3. Click the rectangle and the circle sketches in the Design window.

4. After the preview appears, click the Die Formed option in the dialog box to see how the part would look using that process.

5. Return to the Press Brake option, and set the Chord Tolerance value to 3 mm, as shown in Figure 11.21.

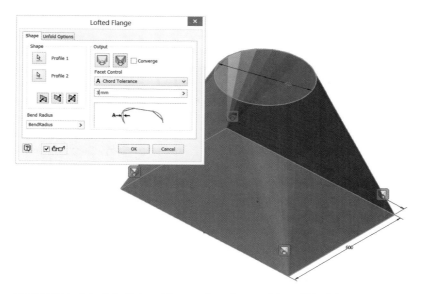

FIGURE 11.21 Being able to see how the transition will be fabricated can help guide decisions.

The model updates to show the number of facets that need to be bent to create the part with this level of precision. Additional options for Facet Control include the angle between facets and the distance between points on the long edge.

6. Click OK to create the feature.

It will not be possible to create a flat pattern of the component as it is now. You will need to create a gap using the Rip feature to be able to unfold the part.

7. Select the Rip tool from the Modify panel.

8. In the dialog box, use the Rip Type drop-down to set the value to Point To Point.

9. Set the model to the Home view, and select the large triangular face aligned with the bottom of the ViewCube® as the rip face.

10. Make the start point the top tip of the face and the end point the middle of the bottom edge, as shown in Figure 11.22.

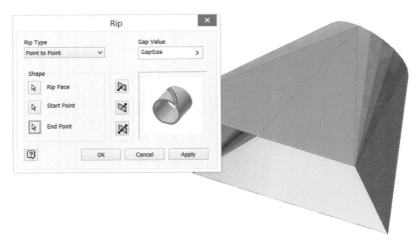

FIGURE 11.22 Making it possible to calculate the flat pattern

11. Click OK to create the rip in the part.

The exercise noted the eventual need to create a flat pattern. Later in this chapter, I will cover the tools for creating a flat pattern.

Working with Existing Designs

If you or your employer has a history of making sheet metal components, it is likely that you already have some flat patterns that work well for your designs. It is possible to use those existing DWG or DXF files to create the 3D part and even to add features to it. These steps show you how to make use of existing AutoCAD flat pattern data.

1. Verify that the 2014 Essentials project file is active, and then open c11-11.ipt from the Parts/Chapter 11 folder.

2. Edit Sketch 1 in the Browser.

3. Click the Insert AutoCAD tool on the Insert panel of the Sketch tab.

4. Open c11-11.dwg from the Parts/Chapter 11 folder.

5. When the Layers And Objects Import Options dialog box opens, click the Next button to keep all layers selected, make sure the Constrain End Points option is selected, and click Finish.

6. When the geometry appears in the sketch, finish the sketch.

7. Restore the Home view so you can see the entire sketch.

8. In the Create panel, click the Face tool, and click all four portions of the sketch.

9. Click the Offset option to change the direction in which the face feature will be developed, and click OK to create the face.

10. In the Browser, expand the face feature, right-click the sketch, and click Share Sketch in the context menu.
   Now the sketch is visible, and you can use the bend lines from the flat pattern to fold the material into a 3D form.

11. Click the Fold tool in the Create panel.

12. For the first Bend Line selection, click the red line beside the number 1. This displays a glyph of a straight arrow showing which side of the bend will be folded and a curved arrow to show the direction of the fold.

13. Click Apply to fold a small portion of the part.

14. For the second edge, click the line next to the number 2, and select Apply.

15. Click the edge near the number 3. Notice that the glyph indicates that it will fold up the wrong portion of the part.

16. Click the Flip Side icon to make sure that the correct geometry (Figure 11.23) will be folded, and click OK.

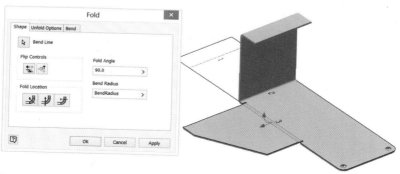

FIGURE 11.23 Folding a flat portion of a face around a line

The component is now suitable for building new drawings or developing new features. Changes to the sketch will modify the folded component as expected.

Adding the Finishing Touches

Creating the big portions of a sheet metal part has been covered so far, but the details are also important. Adding chamfers, fillets, and hems to the metal improves safety for the customer and adds strength to the part. In this exercise you will add the specialized detail features to the part.

1. Verify that the 2014 Essentials project file is active, and then open c11-12.ipt from the Parts/Chapter 11 folder.

2. Zoom in on the top-left corner of the folded part.

3. Start the Corner Chamfer tool from the Modify panel.
   Once the dialog box is open, you can select edges, but unlike the normal Chamfer, this tool only selects the corners on the edge of the sheet so that the feature can be cut from material using the flat pattern.

4. Click the corner shown in Figure 11.24, set the value to 5 mm, and click OK to place the chamfer.

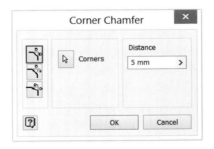 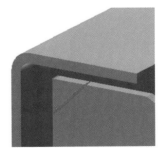

FIGURE 11.24 Corner chamfers can be cut only in the flat pattern.

5. Click the Corner Round tool on the Modify panel.

6. Select the edge shown in Figure 11.25, set the radius to 5 mm, and click OK to place the fillet.

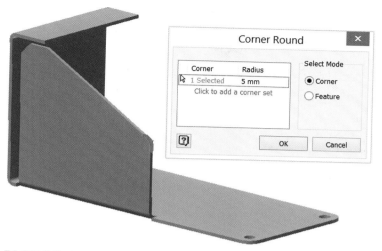

FIGURE 11.25 A corner round follows the same limitations as the corner chamfer.

Another useful option in the Corner Round tool is a feature selection option, which can find all the available edges with one click.

7. Click the Hem tool in the Create panel.

8. After the dialog box opens, click the long edge of the short face (Figure 11.26).

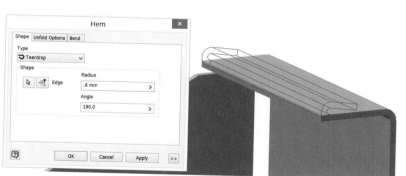

FIGURE 11.26 Hems can be complex features but are added easily.

9. Use the Type drop-down to set the option to Teardrop.

10. Set the Radius value to 0.8 mm, as shown in Figure 11.26, and click OK to place the hem.

The Hem tool's four options should cover almost any requirement for fabrication. Now you can focus on making your components suitable for the manufacturing process.

Preparing the Part for Manufacture

In some cases, you might choose to send data for the finished part to an external resource and allow that person to calculate the material needed to create the part. Inventor has the capability to develop a flat pattern for a part. You can also edit flat patterns to add material that isn't included in the folded part.

Building the Flat Pattern

A flat pattern for a part includes calculations for the stretch of the material as it is formed. The type, thickness, and bend radius can affect this calculation. It is also common for companies to build in extra material to cause deformation, and this is done by modifying the flat pattern, as shown in these steps:

1. Verify that the 2014 Essentials project file is active, and then open c11-13.ipt from the Parts/Chapter 11 folder.

2. Click the Create Flat Pattern tool in the Sheet Metal tab's Flat Pattern panel. Figure 11.27 shows the resulting flat pattern.

FIGURE 11.27 The flat pattern of the part

Take a moment to look at the Browser. The flat pattern appears in the Design window, but the Browser is also updated to show the flat pattern state.

3. Click the Distance tool on the Measure panel of the Tools tab.

4. Click the far-left and far-right edges of the flat pattern, and make note of the measured distance.

5. Return to the Flat Pattern tab, and click Go To Folded Part in the marking menu or the Folded Part panel.

6. Click the Sheet Metal Defaults in from the Setup panel or marking menu, and using the Sheet Metal Rule drop-down, switch to the Steel – 16 ga style.

7. Click OK, and watch the part closely to see it change.
   You should see that the material is thinner and the gaps between edges are smaller, and you will also see that the last flange added has a relief at either end.

8. Return to the flat pattern by double-clicking its icon in the Browser.

9. Make the same measurement as in step 4, and see the difference in the size of the flat pattern.

10. Right-click the flat pattern in the Browser, and click Extents in the context menu. See Figure 11.28 for the results.

Once a flat pattern is generated, you can right-click it in the browser and use the Save Copy As feature to export it as a DXF, DWG, or SAT solid model.

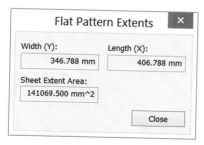

FIGURE 11.28 The Flat Pattern Extents dialog box reports the material needs of the part.

11. Close the dialog box, press the F6 key for a Home view of the flat pattern, and then zoom in on the corner of the part that is near the bottom of the Design window.

12. Start the Fillet tool by pressing F (click OK to modify the flat pattern), and set it to a face fillet.

13. Click the two sides of the part shown in Figure 11.29, and set the radius to 5 mm.

FIGURE 11.29 You can add features like fillets to the flat pattern.

14. Click OK to place the fillet in the flat pattern.

The fillet that was added appears on drawings of the flat pattern but does not propagate back to the 3D model. Features added to the flat pattern are specifically for the use and benefit of the manufacturing process.

Documenting Sheet Metal Parts

The ability to create documentation in the drawing view that you've seen with other solid models works just as well for sheet metal components, but they need even more. For sheet metal, it's important to document not only the folded part but also the flat part, and being able to communicate the process for forming it is a great benefit.

Establishing the Process

In the design phase, the flanges, bends, and faces of the part can be built in any particular order. That's not true for producing the part. Inventor has the ability to specify the order in which you want folds to be made in the solid model, as you'll see in these steps. The order you specify can then be accessed in the drawing.

1. Verify that the 2014 Essentials project file is active, and then open c11-14.ipt from the Parts/Chapter 11 folder.

2. Activate the flat pattern in the model.

3. Select the Bend Order Annotation tool from the marking menu or the Manage panel of the Flat Pattern tab.

    The flat pattern will display numbers that represent the order in which the part would be folded.

4. Click the balloon with the number 6.

5. When the Bend Order Edit dialog box appears, select the Bend Number check box, and change the value to 1. Leave the Unique Number option selected, and click OK.

    This will change the number and the icon to identify it as an override.

6. Change the balloon for 9 to 3, and click OK, as shown in Figure 11.30.

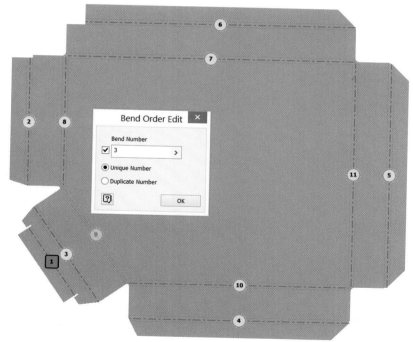

FIGURE 11.30 Like Bill of Material data in the assembly, bend order information is kept in the part file.

Adding this information to the part makes it accessible by anyone who would need to make drawings from your part.

Certification
Objective

Documenting the Process

Sheet metal parts often require two types of drawing. A drawing of the folded part will help quality control and manufacturing verify that the part was built properly. Drawings of the flat patterns can be used to cut the parts out of the raw material. In these steps, you will create and annotate a sheet metal part drawing.

1. Verify that the 2014 Essentials project file is active, and then create a new drawing using ISO.idw from the Metric folder of the New File dialog box.

2. Select the Base View tool from the marking menu or the Create panel of the Place Views tab.

3. Click the Open An Existing File icon, and open the c11-15.ipt file from the Parts/Chapter 11 folder.

4. On the Component tab of the Drawing View dialog box, click the flat pattern for the Sheet Metal view, and set the scale to 1:2.

5. Place the view in the middle of the page, right-click, and choose OK from the context menu.

6. Switch the Ribbon to the Annotate tab, and click Bend on the Feature Notes panel.

7. Click the bend line in the drawing closest to the top; when the note is placed, press the Esc key, or right-click and click OK in the context menu.

    Clicking and dragging the insertion point of the bend note will place the note on a leader.

8. Click the Table tool in the Table panel on the Annotate tab.

9. When the Table dialog box opens, click the drawing view.

    The dialog box changes when the view is selected and lists the columns that will be shown in a bend table.

10. Click OK in the dialog box, and then place the table on the drawing. See Figure 11.31.

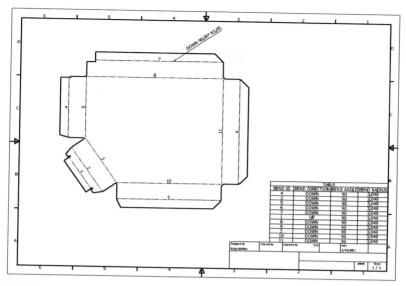

FIGURE 11.31 The bend table and callouts on the drawing of the flat pattern

The numbers of the bends are added to the drawing views and correspond to any edits made to the bend order.

11. Right-click the drawing view, and select Open to view `c11-13.ipt` in its own tab.

12. Right-click the flat pattern in the Browser, and click Save Copy As in the context menu.

13. In the Save Copy As dialog box, use the drop-down menu to change the file type to DXF, and then click Save.

14. In the Flat Pattern DXF Export Options dialog box, set File Version to AutoCAD R12/LT 2 DXF.

15. Activate the Geometry tab, select the Merge Profiles Into Polyline check box, and click OK.

This generates a new DXF file, which is the most common file format for use with two-axis cutting equipment. You can also generate SAT and DWG files for solid export or direct use with AutoCAD.

THE ESSENTIALS AND BEYOND

Sheet metal tools keep the focus on creating the correct geometry as quickly and easily as possible. The use of styles in Inventor to establish the material and its tools that use industry terms makes using these tools more natural for anyone who commonly needs to create sheet metal parts. Tools like Contour Flange can simplify the process by creating many features in one step. Manipulating the flat pattern is a necessity for manufacturing to make sure they can keep using processes that they've used in the past.

ADDITIONAL EXERCISES

▶ Set up templates using the material and sheet gauges you typically use.

▶ Use the Flat Pattern tools on the earlier exercises in the chapter.

▶ Experiment with other Bend options.

▶ Resize and reposition the profiles for the lofted flange and see the update.

CHAPTER 12

Building with the Frame Generator

The specialized tools in Autodesk® Inventor® software help you shortcut arduous and repetitive processes. The design accelerators and plastic part features you worked with in earlier chapters are great examples of this. Another is the Frame Generator. The frames you create with this tool are constructed from dozens of metal shapes contained in the Content Center. Once you place them, you can control how the frame members are joined together, and if the skeleton they're based on changes, the members update as well.

▶ **Creating metal frames**

▶ **Editing metal frames**

Creating Metal Frames

To build a frame, you must have a sketch (2D or 3D) or a solid model to use as a skeleton. You place the frame members by selecting edges or by clicking a beginning point and an endpoint for the frame element. You can tell the frame member how to position itself along its direction based on nine control points around its section that show how the profile will be oriented to the sketch element. You can also offset the profile a distance from the sketch axis or rotate the section.

Beginning the Frame

The techniques for building the skeleton of the frame are no different from any sketching or part modeling technique covered in previous chapters of this book. For that reason, I will shortcut the process by placing a finished skeleton part into an assembly.

1. Make sure that 2014 Essentials is the active project file, and create a new assembly file using the Standard (in).iam template from the English folder under Templates in the New File dialog box.

2. Use the Place tool on the Component panel of the Assemble tab and select the c12-01.ipt file from the Parts\Chapter 12 folder.

3. Right-click when the preview appears and click Place Grounded at Origin from the marking menu to add the frame skeleton, as shown in Figure 12.1.

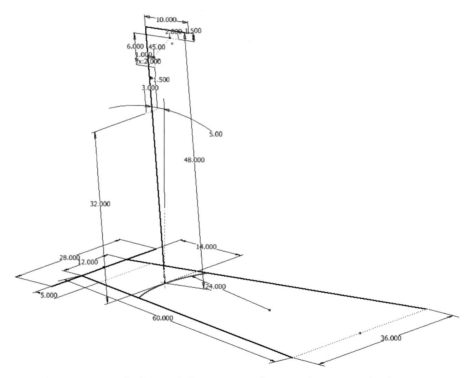

F I G U R E   1 2 . 1   The frame's skeleton consists of a series of parametric sketches.

4. Press the Esc key to quit the Place Component tool.

5. Close the file without saving. For the next exercises, you will use an assembly with the frame already in place.

Whether you load a part file or create a new part in the assembly to use as the skeleton, this part will always be separate from the rest of the frame.

Inserting Members on Edges

The process of placing the members requires that you click the geometry on which you want to place them, select a standard of material, and then click the shape to use and the size. You can also select a color or material property for the member.

 Certification Objective

You will now use a skeleton to begin the process of building your frame:

Insert Frame

1. Make certain that the 2014 Essentials project file is active, and then open the c12-01.iam file from the Assemblies\Chapter 12 folder.

2. Switch the Ribbon to the Design tab, and select the Insert Frame tool from the Frame panel.

3. In the Insert frame dialog box, set Standard to ANSI, Family to ANSI AISC (Square) – Tube, and Size to 3×3×3/16, with the orientation in the center of the profile, as shown in Figure 12.2.

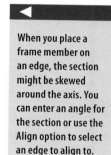

When you place a frame member on an edge, the section might be skewed around the axis. You can enter an angle for the section or use the Align option to select an edge to align to.

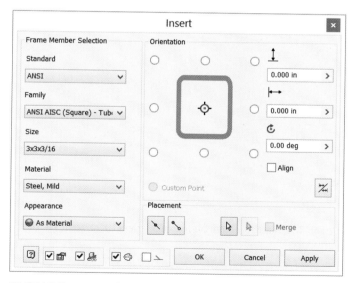

FIGURE 12.2 Setting the type of metal section

4. After setting the proper frame type, deselect the Select Construction option in the Insert dialog box.

   This option makes it possible to toggle whether lines with the Construction override are eligible to have frame members placed on them.

5. Make sure the Insert Members On Edges Placement option is active, and drag a selection window around the entire skeleton. Your results will look like Figure 12.3.

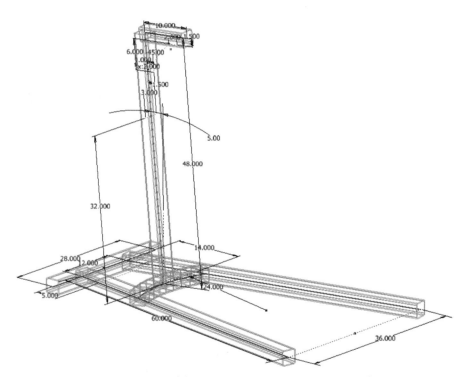

FIGURE 12.3 A preview of the steel components to be placed in the frame

6. Click OK to generate the frame members; then click OK again to approve the creation of the new files and a third time to approve the names of the new files.

   It is possible to change the names and paths of these new files to suit your needs.

Inserting Members between Points

Rather than creating a lot of extra lines or model edges, you can place the members by selecting a start point and an endpoint:

1. Make certain that the 2014 Essentials project file is active, and then open the c12-02.iam file from the Assemblies\Chapter 12 folder.

2. Switch the Ribbon to the Design tab, and select the Insert Frame tool from the Frame panel.

3. In the Insert frame dialog box, set Standard to ANSI, Family to ANSI L (Equal angles) – Angle Steel, and Size to L1.5×1.5×3/16, keeping the orientation in the center of the profile.

4. Make sure the Insert Members Between Points Placement option is active, and click the endpoint of the 3″ line embedded in the upright member of the frame. Then pick the endpoint of the dashed, 12″ pink line at the base of the frame, as shown in Figure 12.4.

When you are working in a production environment, it is a good idea to give unique names to the frame members, subassemblies, and the skeleton when prompted. Filename prompting can also be disabled.

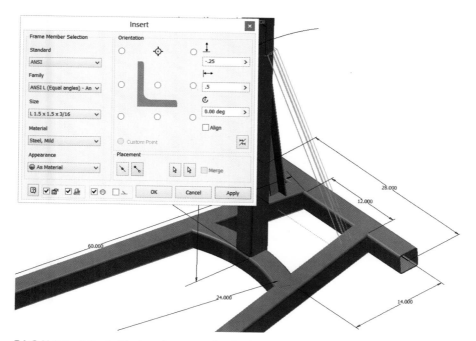

FIGURE 12.4 Placing a frame member by clicking two points

Selection points must be in a sketch. If there is a work point in the model you want to use, you will need to project it into a sketch first.

5. Change the orientation of the profile to the top-center option, as shown in Figure 12.4.

6. Click the Align option and pick an outside edge of the 28″ member. Then set the offset values to -.25 and .5, as shown in Figure 12.4.

7. Click OK to generate the frame member, and then click OK to approve the name of the new file.

The ability to realign the profile gives you flexibility with open metal shapes.

Inserting Members on Curves

Frame members can be placed on curved edges as well. To edit curved frame members, use solid modeling tools like Extrude or Split to remove portions of these components.

1. Make certain that the 2014 Essentials project file is active, and then open the c12-03.iam file from the Assemblies\Chapter 12 folder.

2. Switch the Ribbon to the Design tab, and select the Insert Frame tool from the Frame panel.

3. In the Insert dialog box, set Standard to ANSI, Family to ANSI AISC (Rectangular) – Flat Bar Steel, and Size to 2 1/4×3/16. Set the orientation to the center of the profile.

4. Make sure the Select Construction option is selected in the Insert dialog box to allow selection of the sketch geometry.

5. Make sure the Insert Members On Edges Placement option is active, and click all the red lines and arcs on the skeleton.

6. Change the orientation point to be the middle-left center of the profile, set the Vertical and Horizontal Offset to 0, set the rotation angle of the section to 90 degrees, and select the Merge check box. Your results will look like Figure 12.5.

   You might need to use a different orientation to achieve the correct results. Use the preview as a guide in all situations.

7. After the preview appears and is positioned correctly, click OK to create the new frame member, and click it again to approve the name for the part.

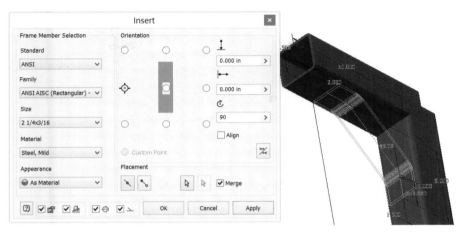

FIGURE 12.5 Building the metal shape across the straight and curved lines

Placing the frame members always follows the same handful of steps. The editing process is also direct. The tools are dialog box–driven and typically require only a couple of selections.

Editing Metal Frames

There are only a few editing tools for the Frame Generator, but they are more than enough to make the type of changes typically needed in the process of designing metal frames. Some of the tools allow for building in gaps for welding operations.

Any change in the skeleton will update the frame members as well as the edits made to them.

Defining Joints with the Miter Tool

A common joint for metal frames is an equal cut on both parts. A gap can be included in the miter equally divided between both parts, or it can be taken from one side or the other. In these steps, you will begin to refine the design of the frame.

1. Make certain that the 2014 Essentials project file is active, and then open the c12-04.iam file from the Assemblies\Chapter 12 folder.

2. Switch the Ribbon to the Design tab, and select the Miter tool from the Frame panel.
   When the Miter dialog box opens, you will be prompted to click two frame members.

3. Click the blue beam for the first option and the red one for the second.

4. Set the gap to 0.130 in, as shown in Figure 12.6, and click OK to create the miter.

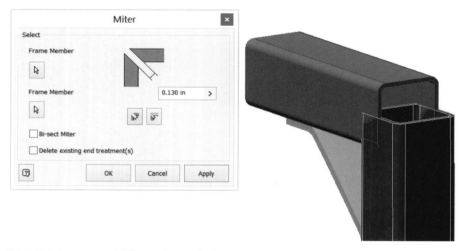

FIGURE 12.6 Adding a miter to the frame

Adding the miter changes the length of both frame members. Any end treatment applied to a frame member can be removed.

Changing an Edit and the Trim To Frame Tool

The Trim To Frame end treatment will take two members and set one to butt against the other and then limit the length of the first member to the width of the second.

The existing frame has a miter applied to two sections. You need to change it to a Trim To Frame joint:

1. Make certain that the 2014 Essentials project file is active, and then open the c12-05.iam file from the Assemblies\Chapter 12 folder.

2. Switch the Ribbon to the Design tab, and select the Trim To Frame tool from the Frame panel.

   When the Trim To Frame dialog box opens, you will be prompted to click two frame members. The first, represented in the dialog box in blue, overlaps the second, represented in yellow.

3. Click the orange section for the first option and the nearly vertical section mitered to it as the second selection, as shown in Figure 12.7.

4. Select the Delete Existing End Treatment(s) check box, and click OK to update the model, as shown in Figure 12.8.

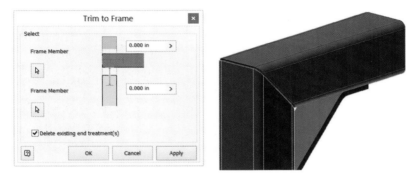

FIGURE 12.7   Modifying the miter to be an overlap

FIGURE 12.8   Trim To Frame will accommodate the sizes of both frame members for a clean overlap.

This exercise removes the miter end treatment and then sets the orange member to overlap the green.

The Trim/Extend Tool

When frame members are placed on the skeleton, they're often overlapping or might fall short of other members they need to butt against. This end treatment allows you to select several members and terminate them at a selected face, whether it means trimming or extending them:

1. Make certain that the 2014 Essentials project file is active, and then open the c12-06.iam file from the Assemblies\Chapter 12 folder.

2. Switch the Ribbon to the Design tab, and select the Trim/Extend tool from the Frame panel.

3. When the dialog box opens, click the two longest members at the base. Once they're selected, right-click and click Continue from the context menu to select the trimming face.

4. Click the face that the two selected segments intersect, as shown in Figure 12.9, and then click OK.

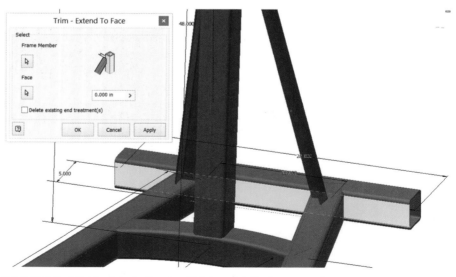

FIGURE 12.9 Select a face to which to match the frame members.

You must select model faces for the trimming surface. The model face can be a face on the skeleton if it is a solid model.

Creating Notches

For more complex joints between frame members, you can cut one member with the shape of another:

1. Make certain that the 2014 Essentials project file is active, and then open the c12-07.iam file from the Assemblies\Chapter 12 folder.

2. Switch the Ribbon to the Design tab, and select the Notch tool from the Frame panel.

   In the Notch dialog box, you select a frame member to cut first and then the member with which to cut it. These will be highlighted in blue and yellow as they are selected.

3. Click the yellow part first and then the red part for the cutting profile, as shown in Figure 12.10.

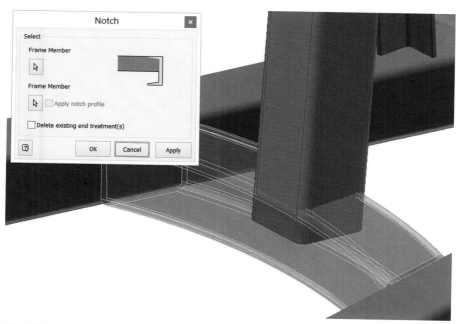

FIGURE 12.10 The notch will update with any changes to the geometry.

4. Click OK to create the feature. See the results in Figure 12.11.

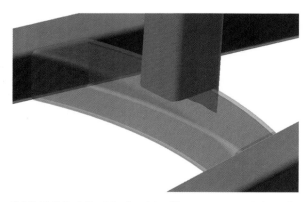

FIGURE 12.11 A notch will remove an exact shape from the selected member.

The Lengthen/Shorten Tool

In cases where a pure notch creates too detailed a cut, you should perform a traditional extruding cut or create a custom coping profile to remove the material based on a specific size and shape.

Lengthen/Shorten

The length of a placed frame member is based on the length of the selected edge or the distance between the selected points. The Trim/Extend tool is useful if you have a face to trim or to which to extend. This tool is useful for changing the length of a member in open space or when you don't want to use a face on another member without changing the skeleton itself. In these steps, you will change the frame member's length beyond the skeleton's original size.

1. Make certain that the 2014 Essentials project file is active, and then open the c12-08.iam file from the Assemblies\Chapter 12 folder.

2. Switch the Ribbon to the Design tab, and select the Lengthen/ Shorten tool from the Frame panel.
   The tool can add length to one end or equally on each end of the selected member.

3. In the Lengthen – Shorten Frame Member dialog box, set the extension length to 4.5 in, and leave the Extension Type set to Lengthen – Shorten Frame Member At One End.

4. For a single direction, the selection is click-sensitive. You need to click the end that will change. Click the end of the yellow piece that is in the open (Figure 12.12), and click OK.

FIGURE 12.12 Selecting the open end of the frame will add length to it.

If you need to add length to an end that has an existing end treatment, you will need to delete the end treatment first or select the Delete Existing End Treatment(s) option.

The Change Tool

Sometimes, you discover that a frame as originally designed needs to be changed. The frame members might need to be lighter or heavier or repositioned. The orientation, size, and family of any frame member can be changed. Using the Change tool, you will select a single or multiple members that are presently the same size. You can then edit the selected members with the same options used to place them:

1. Make certain that the 2014 Essentials project file is active, and then open the c12-09.iam file from the Assemblies\Chapter 12 folder.

2. Switch the Ribbon to the Design tab, and select the Change tool from the Frame panel.

3. When the Change dialog box opens, click the transparent red frame member in the base of the frame. After a short time, its properties will appear in the dialog box.

4. Change the family of the part to ANSI AISC (Rectangular) – Tube and the size to 5×3×3/16 (Figure 12.13); then click OK to update the component.

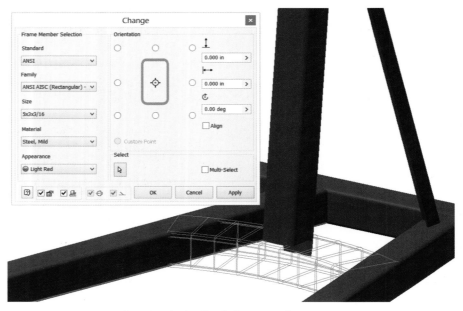

FIGURE 12.13 Changing the family of a frame member

5. Click Yes in the dialog box to approve the change to the family of the component, and click OK to create the new part.

6. Orbit the part to see the effect the change has in the assembly, as shown in Figure 12.14.

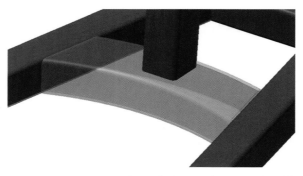

FIGURE 12.14 Edits to frame members are updated by changes to the frame sizes.

The end treatments of the member will be updated even if it's curved.

For all but radical changes to the family of a frame member, the end treatments should be maintained. This is also true for most cases when the skeleton of the frame is changed.

You can also change the standard for a frame element.

Controlling Frame Documentation

Chapters 4 and 8 are centered on the creation of 2D documentation. Each member of a metal frame can be detailed using the same processes described in those chapters, but there is another level of documentation that, while relevant to all assemblies, can be more complex when working with the Frame Generator.

Working with the Bill of Materials and Parts List

When you create a frame using the Frame Generator, a number of steps are automated. When the time comes, you may need to make modifications to get your data presented and documented in the way you need it. In this exercise, you will explore a couple of options for documenting the frame as a whole:

1. Make certain that the 2014 Essentials project file is active, and then open the c12-01.idw file from the Drawings\Chapter 12 folder.

2. Zoom into the view of the assembly and observe the unique numbering of the balloons in the view.

3. Zoom into the parts list in the upper right of the drawing, as shown in Figure 12.15.

You can review adding balloons to the drawing view in Chapter 8, "Creating Advanced Drawings and Annotations."

PARTS LIST			
ITEM	QTY	PART NUMBER	DESCRIPTION
1	63.546 in	AISC - L 1.5 x 1.5 x 3/16 - 31.773	Angle Steel
2	9.325 in	AISC - 2 1/4x5/16 - 9.325	Flat Bar Steel
3	16.000 in	AISC - 3x3x3/16 - 16	Tube
4	46.500 in	AISC - 3x3x3/16 - 46.5	Tube
5	28.000 in	AISC - 3x3x3/16 - 28	Tube
6	117.551 in	AISC - 3x3x3/16 - 58.775	Tube
7	18.747 in	AISC - 5x3x3/16 - 18.747	Tube
8	1	c12-02	

FIGURE 12.15 The parts list displays the structure of the assembly, but this structure can be manipulated.

The parts list uses a basic grouping structure that is shown in the QTY column, where you see the total length of some of the components combined when there are two components of the same length.

4. Right-click on the parts list and select Bill Of Materials from the context menu.

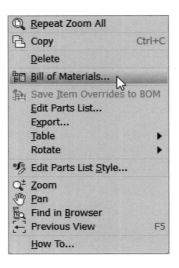

The bill of materials is held within the assembly. When you make changes to the structure, naming, or any other property in the Bill Of Materials dialog, you are making a change to the assembly.

5. In the Bill Of Materials dialog box, expand the structure of the Frame c12-10 component.

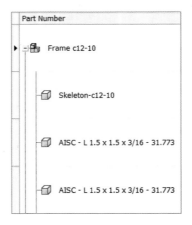

Next you will modify the part number of the individual components using properties of the bill of materials so you don't have to manually change each one.

To save time and add consistency when creating drawings, create a template based on an assembly with the proper bill of materials (BOM) elements using Application menu ➤ Save Copy As ➤ Save Copy As Template.

6. Pick the Part Number cell of the first component below the Skeleton part (probably AISC - L 1.5×1.5×3/16 - 31.773).

7. Click the Create Expression tool in the upper left of the dialog box.

8. Highlight the existing text and use the Backspace key to delete it.

9. Use the Property pull-down to locate the Description property and click the Insert Property button to the right to add the value to the field, as shown in Figure 12.16.

FIGURE 12.16 Building an expression to name parts based on properties

10. Press the spacebar, add a hyphen and another space, and then use the Property pull-down to add the Stock Number property.

11. Click OK to create the new naming convention for your parts in the bill of materials.

12. With the Part Number cell updated, click and drag the small square in the lower-right corner of the cell to copy this expression to the other members of the frame.

An additional part is shown beneath the members of the frame. This is the component that contains the sketch the frame is built on. This component does not need to be included in a parts list, so you need to tell the bill of materials that it is to be ignored by a parts list.

13. Double-click in the cell in the BOM Structure column of the 12-02 .ipt row to activate a pull-down menu.

14. Use the pull-down menu to set the structure to Phantom.

If you look at these new descriptions, you will notice that some of them have the same description. You can use the bill of materials structure to improve how this is handled.

15. In the Bill Of Materials dialog box, select the Structured tab to see the updated structure.

In this assembly, there are two groups of components that have the same descriptions. The first set is Tube - 3×3×3/16. The length is displayed as *Varies* because the components have been grouped but are not all the same length. The second is Angle Steel - L 1.5×1.5×3/16. The Unit QTY is listed as a total because the components are the same length. In the QTY column you will see the total length of the material used for the two groups, as shown in Figure 12.17.

16. Click Done to close the Bill Of Materials dialog box and update the drawing.

The parts list will update showing the new part numbers and summarized QTY values as well as updated item numbers. The balloon numbers have been updated as well. There are now redundant balloons because the balloons refer to a material rather than a component. You will make another modification to the bill of materials to allow better representation of the components in the assembly.

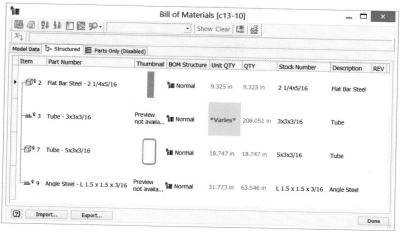

FIGURE 12.17 The bill of materials can be controlled to greatly affect how components are grouped.

17. Right-click on the parts list and select Bill Of Materials to return the dialog box to the screen again.

18. Click on the Part Number Row Merge Settings button in the top center of the dialog box.

19. Deselect the Enable Row Merge On Part Number Match check box.

20. Click OK to update the bill of materials.

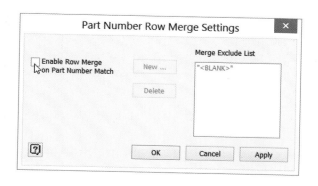

21. Click the Sort Items button near the top left of the Bill Of Materials dialog box.

22. When the Sort dialog box opens, choose Part Number in the Sort By pull-down and Unit QTY in the Then By pull-down. Then click OK to update the bill of materials.

23. Click the Renumber Items button next to the Sort Items button.

24. Set the Start value to 1 and make sure the Increment value is 1; then click OK to renumber all of the components based on the order set in using the Sort tool.

25. Click Done to close the dialog box.

    The parts list will be updated but still may not display exactly like the bill of materials. This is further evidence that the parts list is a separate tool that draws information but is not controlled by the bill of materials.

    Now you will make a change to the parts list to alter its display; this change doesn't affect the bill of materials.

26. Right-click on the parts list and select Edit Parts List from the context menu.

27. Click the Group Settings button to open the Group Settings dialog box.

28. Select the Group option, and from the First Key pull-down select Browse Properties.

29. When the Parts List Column Chooser dialog box opens, select UNIT QTY from the list of Available Properties and click OK.

30. In the Group Settings dialog box, deselect the Display Group Participants check box.

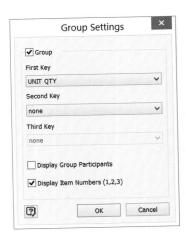

31. Click OK to close the dialog box and update the information in the Parts List dialog box.

32. Click OK to close the Parts List dialog box and to update the parts list, as shown in Figure 12.18.

PARTS LIST			
ITEM	QTY	PART NUMBER	DESCRIPTION
1, 2	63.546 in	Angle Steel - L 1.5 x 1.5 x 3/16	Angle Steel
7, 8	117.55 in	Tube - 3x3x3/16	Tube
3	9.325 in	Flat Bar Steel - 2 1/4x5/16	Flat Bar Steel
4	16.000 in	Tube - 3x3x3/16	Tube
5	28.000 in	Tube - 3x3x3/16	Tube
6	46.500 in	Tube - 3x3x3/16	Tube
9	18.747 in	Tube - 5x3x3/16	Tube

FIGURE 12.18 Restructuring and editing can be done to the Parts List to add detail beyond the bill of materials.

This may not be the only way of displaying the parts list that you would choose. There are many options for showing how individual components made from a standard material and combined in a frame can be shown in a drawing to help manufacturing understand the needs of the design.

Changing the Frame Skeleton

The great advantage of using the Frame Generator over creating individual parts based on extruded profiles is the association between the frame members and the skeleton of the frame. A change to the skeleton updates the members of the frame rather than requiring you to edit individual parts one at a time.

1. Make certain that the 2014 Essentials project file is active, and then open the c12-02.idw file from the Drawings\Chapter 12 folder.

Observe the parts list, which reflects the quantity of material used in the frame as it is. Presently there are 63.5″ of angle steel material and 208.1″ of 3 × 3 tube used in the assembly.

2. Right-click in the drawing view and select Open from the context menu.

This will open the assembly that the drawing was created from rather than you having to use File ➢ Open.

3. When the assembly opens, double-click the 48.000 dimension to activate the sketch.

4. Double-click that dimension again to edit its value and change it to 54, and then click the check mark to update the dimension.

5. In Edit the angle Dimension changing the angle from 5 deg to 10 deg (Figure 12.19), and click the check mark to complete the edit.

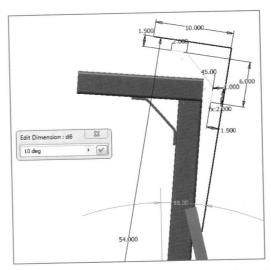

FIGURE 12.19 Editing the parametric dimensions of the skeleton

6. Right-click and select Finish Edit from the marking menu.
   This will update the assembly and change the length of some components while not affecting others.

7. Use the tabs at the bottom of the graphics window to change back to the drawing and see the change reflected in the parts list, as shown in Figure 12.20.

PARTS LIST		
QTY	STOCK NUMBER	DESCRIPTION
9.3 in	2 1/4x5/16	Flat Bar Steel
18.7 in	5x3x3/16	Tube
61.8 in	L 1.5 x 1.5 x 3/16	Angle Steel
214.1 in	3x3x3/16	Tube

FIGURE 12.20 The component lengths update as a result of the change to the assembly.

The simplicity of the Frame Generator tools and the associativity of the generated frame with the skeleton make metal frames easy to create and maintain. If you have experience creating these types of frames, the tools should help you be more productive immediately.

THE ESSENTIALS AND BEYOND

The Frame Generator tools open a new way of constructing the frames that are so common in machine design. Placing the frame members directly from the Content Center saves you the work of having to sketch the shapes and sizes of the parts. Using tools such as Trim/Extend, Trim To Frame, and Miter can save you hours of editing on a simple change. Using this workflow also bypasses the placement of the assembly constraints while maintaining the advantages of separate parts for detailing needs.

ADDITIONAL EXERCISES

▶ Make other changes to the skeleton model to see the effects.

▶ Try changing components to a non-ANSI standard shape.

▶ Change the material of the frame members to see how it affects the mass properties of the assembly.

▶ Experiment with changing closed profile members to ones with open profiles and see how the end treatments behave.

CHAPTER 13

Working in a Weldment Environment

A weldment is a specialized assembly. Like the sheet metal tools, the tools that the Autodesk® Inventor® software provides for working with weldments are built around the workflows and demands of a production environment. A weldment is an assembly that is regarded as a single, inseparable part.

Building the weldment as an assembly first makes it easy to understand how the components need to be positioned and what components are included in the weldment. The weldment environment also understands the ramifications weldments have on Bill of Materials (BOM) structure and how to handle the weldment as a single component and document its members when needed.

Welds that are added to the weldment exist only in the environment and do not require you to maintain more files.

▶ **Converting an assembly**

▶ **Calculating a fillet weld**

▶ **Preparing to apply weld features**

▶ **Applying weld features**

▶ **Adding machined features to the weldment**

▶ **Documenting welds and weldments**

Converting an Assembly

There are a few special environments or shells within an assembly, and in order to access the weldment tools for Inventor, you must convert your assembly to a weldment. To do so, follow these steps:

Certification Objective

1. Make certain that the 2014 Essentials project file is active, and then open c13-01.iam from the Assemblies\Chapter 13 folder.

2. Click the Convert To Weldment tool on the Convert panel of the Environments tab.

3. Click Yes when you see the message telling you that you will not be able to revert this data to a basic assembly.

   After you confirm the conversion, the Convert To Weldment dialog box appears.

4. Set Weld Bead Material to Steel, Mild, Welded using the drop-down menu.

5. Click OK to add the Weld tab to the Ribbon (Figure 13.1), and populate the Browser with a Preparations, Welds, and Machining group.

FIGURE 13.1 The three additional states of a weldment

> Once converted, a weldment cannot be returned to a regular assembly.

The assembly is still an assembly; it will just be treated differently in the BOM and have access to new tools.

Calculating a Fillet Weld

The design accelerators covered in Chapter 9, "Advanced Assembly and Engineering Tools," focused on tools that generated components. There is another class of tools associated with design accelerators: the calculators. The Inventor Weld Calculator tools can calculate several types of welds and solder joints. The following steps show an example of how you can use the calculator to remove guess work.

1. Make certain that the 2014 Essentials project file is active, and then open c13-02.iam from the Assemblies\Chapter 13 folder.

2. Expand the Weld Calculator tool in the Weld panel on the Weld tab.

3. Click the Fillet Weld Calculator (Plane) tool.

   The goal is to keep the weld at the base of the blue arm less than 7 mm. The calculator helps you determine whether this is reasonable.

4. Set the weld form to an all-around rectangle.

5. Set the weld loads to be the bending force parallel with the neutral axis of the weld group.

6. Set the dialog box options as follows:

   ▶ Bending Force Value = 12000 N

   ▶ Force Arm length = 365 mm

   ▶ Weld Height = 5 mm

   ▶ Weld Group Height = 70

   ▶ Weld Group Width = 70

   ▶ Joint Material = Structural Steel SPT360

7. Click Calculate to get the results.

   The calculator tells you that with the default safety factor of 2.5, you would need a weld bead of just over 5 mm, but the stresses are too high.

8. Change the Weld Height value to 6 mm, and click Calculate again to find a weld size that will work. Figure 13.2 shows the results.

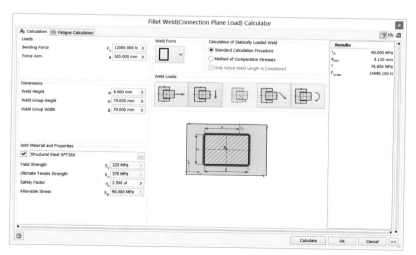

FIGURE 13.2 The properties for calculating the correct fillet weld size

Now that you've calculated the weld, you can apply it with confidence that it is sound. This and the other weld calculators can be used at any time to create a new weld or guide your edits to an existing weld.

Preparing to Apply Weld Features

Preparations are features that are applied to a part after it has been cut but before the weld is applied. These features do not appear in the source part file, because a part might be placed in different assemblies and need different treatments for each. A preparation can be an extrusion, revolved feature, hole, fillet chamfer, or sweep. The most common is the chamfer. In these steps, you will modify a component in the assembly as it would be done in the shop.

1. Make certain that the 2014 Essentials project file is active, and continue using or open c13-02.iam from the Assemblies\ Chapter 13 folder.

Preparation

2. To access the Preparation tools, click the Preparation tool in the marking menu or the Process panel of the Weld tab.

   This step makes tools available in the Ribbon and also changes the Browser to show that these changes are added under the Preparations folder, not to the components in the assembly.

Chamfer

3. Click the Chamfer tool on the Weld tab in the Preparation And Machining panel.

4. Set the distance to 3 mm, and click the three edges highlighted in Figure 13.3.

5. Click OK to add the chamfers to the components.

6. Select Return from the Ribbon to leave the Preparation tools. Figure 13.4 shows the results.

Now that you've added the preparation features you need, you can begin to add some welds to the assembly.

F I G U R E   1 3 . 3   Adding a chamfer to edges to prepare for the weld

F I G U R E   1 3 . 4   The chamfers appear
in the weldment but not the individual parts.

Applying Weld Features

Three types of welds are available: fillet, groove, and cosmetic. In the next set of exercises, you will apply all of them and use a couple of options as well.

Adding a Fillet Weld and the End Fill Tool

Fillet welds are applied by selecting a minimum of two faces. There are two selection sets, but you can add as many faces to each set as you need to define the feature:

Welds

Fillet

1. Make certain that the 2014 Essentials project file is active, and then open c13-03.iam from the Assemblies\Chapter 13 folder.

2. Access the weld tools by clicking the Welds icon in the Process panel on the Weld tab or the marking menu.

3. When the tools appear in the Weld panel of the Weld tab marking menu, click the Fillet tool.

4. In the Fillet Weld dialog box, select the Chain check box, and then click the exterior surface of the blue tube.

5. Click the icon for selection set 2, and then select the face on the yellow plate against which the tube butts.
   A preview of the fillet weld appears at the default value of 10 mm or the last size you used.

6. Set the first fillet value to 6 mm per the weld calculation done earlier in this chapter.

7. Click Apply to generate the fillet weld.

8. Deselect the Chain option.

9. For the first selection, click the face on the blue tube adjacent to the green gusset.

10. Switch to the second selection set, and click the triangular side and the chamfer surface shown in Figure 13.5.

11. Click OK to place the weld.
    Because you selected the side and chamfer face on the gusset, the weld fills in all of the space. Selecting only the blue tube and the side of the gusset could still generate a weld, but it would leave a space.

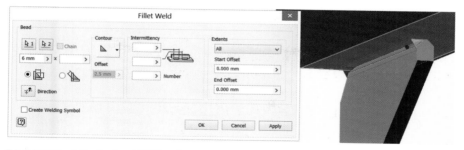

FIGURE 13.5 A weld filling the gap left by a chamfer

One last touch is needed to make the weld look better.

12. In the Weld panel of the Weld tab, click the End Fill tool.
The bare ends of the weld bead should highlight. If they do not, click them.

13. Click one of the ends of the bead, and an image will be added to it, as shown in Figure 13.6.

FIGURE 13.6 Adding an end fill improves the appearance of the welds.

14. Press the Esc key to end the End Fill tool.

Now you've added fillet welds to the assembly. These welds do not require any new part files to maintain them in the assembly.

Adding a Groove Weld

Groove welds are used for filling gaps between parts. The direction of the gap to be closed is established in the dialog box. There is another interesting option on which this exercise will focus — the radial gap. In these steps, you will create two different types of Groove weld.

1. Make certain that the 2014 Essentials project file is active, and then open c13-04.iam from the Assemblies\Chapter 13 folder.

2. Access the weld tools by clicking Welds in the marking menu or in the Process panel on the Weld tab.

Groove

3. Click the Groove tool in the Weld panel.

4. Click the exterior cylindrical face of the red part for Face Set 1.

5. Click the selection icon of Face Set 2, and then click the cylindrical face of the hole in the blue part around the red part.

6. Select the Radial Fill check box in the Fill Direction group. This will present the preview shown in Figure 13.7 of a weld closing the gap between the parts.

FIGURE 13.7 The preview of the groove weld

7. Click OK to create the groove weld.

The groove weld could be left as is, or it could be added as part of a manufacturing process that would now add another weld to it.

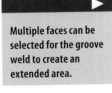

Multiple faces can be selected for the groove weld to create an extended area.

Adding a Cosmetic Weld and Weld Symbols

The fillet and groove are the most common welds and therefore generate physical geometry to represent them, but there are many types of welds that need to be documented and considered in the weldment. A cosmetic weld will apply a weld to an assembly but will not build the geometry itself, as you will see in these steps:

1. Make certain that the 2014 Essentials project file is active, and then open `c13-05.iam` from the `Assemblies\Chapter 13` folder.

2. Access the Weld tools by clicking Welds in the marking menu or in the Process panel on the Weld tab.

3. In the Weld panel of the Weld tab or from the marking menu, select the Cosmetic tool.

4. In the cosmetic weld, set the Area value to 6 mm^2.

5. Leave the Extents set to All, and select the Create Welding Symbol check box.

    This expands the dialog box and allows you to select the specific type of weld that the cosmetic weld will represent. The default is a fillet weld for the opposite side.

6. Click the Swap Arrow/Other Symbols icon to make the fillet weld for the Arrow side.

7. Make sure the Bead selection icon is active, and click the edge shown in Figure 13.8.

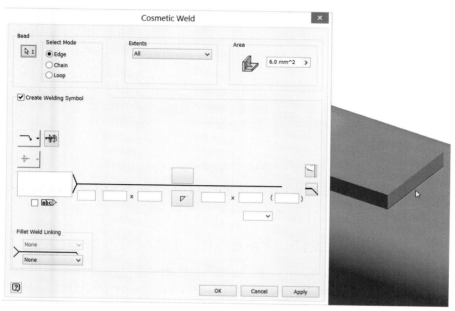

FIGURE 13.8 Select the edge for the cosmetic weld.

8. Click OK to place the weld, which is represented by a highlighted edge and the weld symbol shown in Figure 13.9.

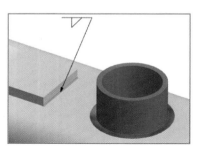

FIGURE 13.9 A weld symbol can be added to any weld placed.

T I P   You can set the weld symbol inside the model. Move near the midpoint of the highlighted edge to show the concealed symbol, and use the grip points to relocate it.

Adding an Intermittent Fillet Weld

Sometimes referred to as a *stitch weld*, this feature creates fillet welds of a set length and distance apart, with the ability to specify a limited number. Follow these steps to learn how to set up this special feature:

1. Make certain that the 2014 Essentials project file is active, and then open c13-06.iam from the Assemblies\Chapter 13 folder.

2. Access the weld tools by clicking the Welds icon in the Process panel on the Weld tab or in the marking menu.

3. When the tools appear in the Weld panel of the Weld tab, click the Fillet tool.

4. When the Fillet Weld dialog box opens, set the first fillet size to 6 and the second to 4.

5. Change the contour from Flat to Convex, and set the offset to 1.

6. Set the top Intermittency value to 9 and the one below it to 10.

7. Click the first bead selection on the side of the block and the second on the top face of the tube, as shown in Figure 13.10.

8. Click OK to place the fillet welds.

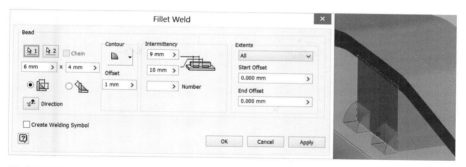

FIGURE 13.10 A preview of the intermittent fillet weld

Now you've added the final fillets to this weldment, and what was an assembly of parts can be produced as a single deliverable.

Adding Machined Features to the Weldment

Because of the distortion from welding components together and the need for some features to pass through several of the welded components, sometimes it is best to just machine features after components are welded together:

1. Make certain that the 2014 Essentials project file is active, and then open c13-07.iam from the Assemblies\Chapter 13 folder.

2. Access the machining tools by clicking the Machining icon in the Process panel of the Weld tab or in the marking menu.
   There is already a sketch in the assembly; it was added in the machining process and appears in the Browser that way.

Machining

3. Click the Extrude tool in the Preparation And Machining panel.

4. Use the Extents drop-down in the Extrude dialog box to set its value to To.

5. Click the interior surface of the red part for the termination. The preview will look like Figure 13.11.

6. Click OK to remove all material between the sketch plane and the selected face.

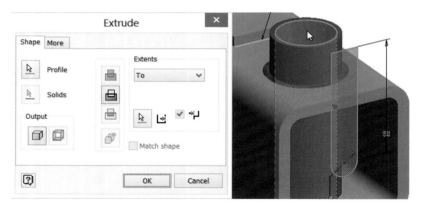

FIGURE 13.11 Adding a postweldment machined feature

Like preparation features, the cuts made by the extrusion do not appear in the source files. They remove the original component geometry and the geometry created by the welds themselves.

Documenting Welds and Weldments

Inventor keeps track of the physical properties of the welds as they're added. It also keeps track of the various states of the model so that the preparation, welds, and machining geometry can be documented as a workflow.

Extracting the Physical Properties of the Beads

To help estimate the materials needed to create a weldment, you have the BOM for the components, but you also need the volume and mass of the weld beads. Follow these steps to find how much mass the welds add to the assembly.

1. Make certain that the 2014 Essentials project file is active, and then open c13-08.iam from the Assemblies\Chapter 13 folder.

2. Expand the Welds category in the browser, and right-click the Beads folder.

3. Select iProperties from the context menu.

4. Switch to the Physical tab, and click Update to get the mass and volume information on the weld material for the assembly. See Figure 13.12.

FIGURE 13.12 The mass and volume of the welds are calculated even for the cosmetic welds.

You can also get the properties for individual beads by generating a bead report.

5. Close the iProperties dialog box, and click the Bead Report tool in the Weld panel on the Weld tab.

6. When the weld bead report is opened, click Next.

7. Choose a name and destination for the XLS file. Figure 13.13 shows an example.

Document		ID	Type	Length	UoM	Mass	UoM	Area	UoM	Volume	UoM
C:\Inventor 2014 Essentials\Assemblies\Chapter 13\c13-08.iam											
		Fillet Weld 1	Fillet	105.872	mm	0.011	kg	2.12E+03	mm^2	1.41E+03	mm^3
		Fillet Weld 2	Fillet	259.398	mm	0.038	kg	5.59E+03	mm^2	4.90E+03	mm^3
		Fillet Weld 3	Fillet	109.196	mm	0.011	kg	2.04E+03	mm^2	1.34E+03	mm^3
		Fillet Weld 4	Fillet	109.196	mm	0.011	kg	2.04E+03	mm^2	1.34E+03	mm^3
		Groove Weld 1	Groove	N/A		0.017	kg	2.12E+03	mm^2	2.19E+03	mm^3
		Cosmetic Weld 1	Cosmetic	30	mm	0.001	kg	6	mm^2	180	mm^3
		Fillet Weld 5	Fillet	28	mm	0.002	kg	383.879	mm^2	303.85	mm^3
		Fillet Weld 6	Fillet	28	mm	0.002	kg	383.879	mm^2	303.85	mm^3

FIGURE 13.13   The physical properties of each weld

TIP   Even though there is no physical geometry, a cosmetic weld is capable of contributing to the mass calculations if you apply an Area value in the dialog box.

Creating Weldment Drawings

The purpose of a weldment environment is to accommodate the needs of the production workflow. Documenting the steps is also important. Inventor can create drawing views of the stages of the weldment, as you will see next:

1. Make certain that the 2014 Essentials project file is active, and then create a new drawing using the ISO.idw template under the Metric folder under Templates in the New File dialog box.

Base

2. Create a base view from the marking menu or the Create panel of the Place Views tab.

3. Click the Open Existing File button, browse to Assemblies\Chapter 13, and click the c13-08.iam file.

4. Set the orientation of the view to Iso Bottom Left and the scale to 1:3.

5. Click the Shaded style.

6. Switch to the Display Options tab, and uncheck Tangent Edges.

7. Change to the Model State tab in the dialog box.

8. In the Weldment group, set the type of view you want to place to Assembly, and then click a location on the drawing for the view, as shown in Figure 13.14.

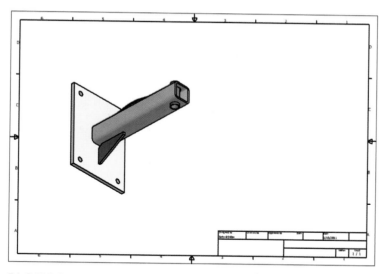

FIGURE 13.14 A drawing view of the Assembly state

9. Start the Base view tools again.

10. Repeat steps 3 through 7.

11. Set the view type for the weldment group to Machining, and place the view so that your drawing looks like Figure 13.15.

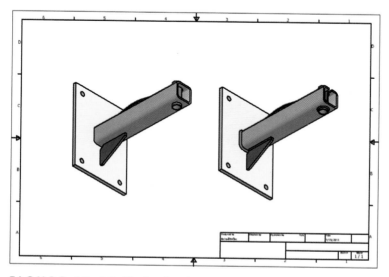

FIGURE 13.15 The drawing showing the beginning and end of the process

The intermediate steps of preparations can be shown in the drawing as well, and any type of drawing can be created from these states, just as you would any other file.

THE ESSENTIALS AND BEYOND

Working with tools whose approach is based on the production process builds efficiency into the design from the beginning and makes it easier to edit should production require that changes be made to the weldment.

The separate environments within a weldment make it easier to keep the focus on the steps of manufacturing a welded component and making sure that each step has all the information needed to complete the task.

By offering the flexibility of the cosmetic weld in addition to welds that add physical geometry to the weldment, like fillet and groove, Inventor ensures that you'll be able to document and plan for any type of welding that your design needs.

ADDITIONAL EXERCISES

▶ Try using a groove weld to connect the gusset to the tube instead of the fillet weld.

▶ See how different load types affect the results of the weld calculator.

▶ Add additional types of features to the machining state.

▶ Add welding symbols to the drawing from the Symbols panel of the Annotate tab.

Autodesk Inventor Certification

Autodesk certifications are industry-recognized credentials that can help you succeed in your design career, providing benefits to both you and your employer. Getting certified is a reliable validation of skills and knowledge, and it can lead to accelerated professional development, improved productivity, and enhanced credibility.

This Autodesk Official Training Guide can be an effective component of your exam preparation. Autodesk highly recommends (and we agree!) that you schedule regular time to prepare by reviewing the most current exam preparation road map available at www.autodesk.com/certification, using Autodesk Official Training Guides like this one, taking a class at an Authorized Training Center (find them near you here: www.autodesk.com/atc), and using a variety of other resources — including plenty of hands-on experience.

To help you focus your studies on the skills you'll need for these exams, the following tables show objectives that could potentially appear on an exam and in which chapter you can find information on that topic — and when you go to that chapter, you'll find certification icons like the one in the margin here.

Certification Objective

Table A.1 provides information for the Autodesk Certified User exam and lists the section, exam objectives, and chapter where the information is found. Table A.2 provides information for the Autodesk Certified Professional exam. The sections and exam objectives listed in the table are from the Autodesk Certification Exam Guide.

These Autodesk exam objectives were accurate at press time; please refer to www.autodesk.com/certification for the most current exam road map and objectives.

Good luck preparing for your certification!

T A B L E   A . 1   Autodesk Certified User exam sections and objectives

Topic	Learning Objective	Chapter
User Interface: Primary Environments	Name the four primary environments: Parts, Assemblies, Presentations, and Drawings.	Chapter 1
User Interface: UI Navigation/ Interaction	Name the key features of the user interface (Ribbon ➤ panels ➤ tabs).	Chapter 1
	Describe the listing in the Browser for an assembly file.	Chapter 1
	Demonstrate how to use the context (right-click) menus.	Chapter 1
	Demonstrate how to use the menus.	Chapter 1
	Demonstrate how to add Redo to the Quick Access toolbar.	Chapter 1
User Interface: Graphics Window Display	Describe the steps required to change the background color of the graphics window.	Chapter 1
	Use Application Options ➤ Display.	Chapter 1
	Demonstrate how to turn on/off the 3D Indicator.	Chapter 1
User Interface: Navigation Control	Describe the functionality of the ViewCube®.	Chapter 1
	Describe the Navigation bar.	Chapter 1
User Interface: Navigation Control	Name the navigation tools started by the F2 to F6 shortcut keys: Pan (F2), Zoom (F3), Free Orbit (F4), Previous View (F5), Home View (F6).	Chapter 1
File Management: Project Files	Name the file extension of a project file (.ipj).	Chapter 1
	List the types of project files that can be created.	Chapter 1
	Define the term workspace.	Chapter 1
	List the types of files stored in a library.	Chapter 1
	List the three categories in Folder Options.	Chapter 1
	Describe how to set the active project.	Chapter 1

Sketches: Creating 2D Sketches	Name the file extension of a part file (.ipt).	Chapter 1
	Describe the purpose of a template file in the sketch environment.	Chapter 1, Bonus Chapter 1
	Describe the function of the 3D Coordinate System icon.	Chapter 1
	Define a sketch plane.	Chapters 2, 6
	Label the entries on the Browser.	Chapter 1
Sketches: Draw Tools	Complete a 2D sketch using the appropriate draw tools: Line, Arc, Circle, Rectangle, Point, Fillet, Polygon.	Chapters 2, 6, 7
Sketches: Sketch Constraints	List the available geometric constraints: coincident, colinear, concentric, fixed, parallel, perpendicular, horizontal, vertical, tangent, symmetric, and equal.	Chapter 2
	Describe parametric dimensions (general and automatic).	Chapter 2
	Describe how to control the visibility of constraints.	Chapter 2
	Describe the degrees of freedom on a sketch and how they can be displayed.	Chapter 2
Sketches: Pattern Sketches	Demonstrate how to pattern a sketch (rectangular, circular, and rotate).	Chapter 2
Sketches: Modify Sketches	Demonstrate how to move a sketch.	Chapter 2
	Demonstrate how to copy a sketch.	Chapter 2
	Demonstrate how to rotate a sketch.	Chapter 2
	Demonstrate how to trim a sketch.	Chapter 2
	Demonstrate how to extend a sketch.	Chapter 2
	Demonstrate how to offset a sketch.	Chapter 2

(Continues)

TABLE A.1   *(Continued)*

Topic	Learning Objective	Chapter
Sketches: Format Sketches	Describe how to format sketch linetypes.	Chapters 2, 7
	Discuss overconstrained sketches.	Chapter 2
Sketches: Sketch Doctor	Examine a sketch for errors.	Chapter 2
Sketches: Shared Sketches	Describe the function of a shared sketch.	Chapters 2, 3, 6, 7, 9, 10
Sketches: Sketch Parameters	Describe how parameters define the size and shape of features.	Chapter 2
Parts: Creating Parts	Name the file extension of a part file (.ipt).	Chapter 1
	Label the entries on the Browser.	Chapter 1
	Define a base feature.	Chapter 3
	Define an unconsumed sketch.	Chapter 3
	Demonstrate how to create an extruded part.	Chapters 2, 3, 6, 7
	Demonstrate how to create a revolved part.	Chapters 3,6, 7
	Demonstrate how to create a lofted part.	Chapter 7
	Describe the termination options for a feature.	Chapters 3, 6, 7
	Demonstrate how to create a hole feature.	Chapters 3, 7
	Demonstrate how to create a fillet feature.	Chapters 3, 7
	Demonstrate how to create a chamfer feature.	Chapter 3
	Demonstrate how to create a shell feature.	Chapter 7
	Demonstrate how to create a thread feature.	Chapter 7
Parts: Work Features	Describe the use of work features (plane, point, axis) in the part creation workflow.	Chapter 6

Parts: Pattern Features	Demonstrate how to create a rectangular pattern.	Chapter 7
	Demonstrate how to create a circular pattern.	Chapter 7
	Demonstrate how to mirror features.	Chapter 7
Parts: Part Properties	Describe part properties and how they are applied.	Chapter 2, Bonus Chapter 1
Assemblies: Creating Assemblies	Name the file extension of an assembly file (`.iam`).	Chapters 1, 4, 8, 10, 11, Bonus Chapter 2, Bonus Chapter 3
	Label the entries on the Browser.	Chapter 1
	Name the six degrees of freedom on a component.	Chapter 5
	Demonstrate how to place a part in an assembly.	Chapter 5
	Discuss degrees of freedom and a grounded part.	Chapter 5
	Demonstrate how to apply various assembly constraints.	Chapters 5, 9
	Describe the various assembly environment techniques (top-down, bottom-up, and middle-out).	Chapters 5, 9
	Demonstrate how to create a new part in the assembly environment.	Chapter 9
	Demonstrate how to place a Content Center part in an assembly.	Chapters 5, 9
Assemblies: Animation Assemblies	Demonstrate how to animate an assembly using drive constraints.	Chapter 9
Assemblies: Adaptive Features, Parts, and Subassemblies	Demonstrate how to make and use an adaptive part.	Chapter 9

(Continues)

TABLE A.1   *(Continued)*

Topic	Learning Objective	Chapter
Presentations	Name the file extension of a presentation file (. ipn).	Chapter 1, Bonus Chapter 2, Bonus Chapter 4
	Label the entries on the Browser.	Chapter 1
	Discuss the various uses of Presentation files.	Chapter 8, Bonus Chapter 4
	Demonstrate how to apply tweaks to a part.	Bonus Chapter 4
	Demonstrate how to apply trails to a part.	Bonus Chapter 4
	Demonstrate how to animate an assembly.	Bonus Chapter 4
Drawings	Name the file extension of a drawing file (. idw).	Chapters 1, 4, 8, 12, Bonus Chapter 1, Bonus Chapter 2, Bonus Chapter 3
	Describe the use of template files.	Chapter 1, Bonus Chapter 1
	Label the entries on the Browser.	Chapter 1
	Describe the content within Drawing Resources.	Bonus Chapter 1
	Demonstrate how to create a part drawing.	Chapters 4, 8, 11
	Demonstrate how to create an assembly drawing.	Chapter 8
	Describe the various annotation options.	Chapters 4, 8, 11
	Demonstrate how to add balloons to an assembly.	Chapter 8

Sheet Metal: Creating Sheet Metal Parts	Name the file extension of a sheet metal part file (.ipt).	Chapter 11
	Discuss the use of sheet metal defaults.	Chapter 11
	Demonstrate the creation of a sheet metal bend. Use Create Tools ➤ Bend, Face, and Flange.	Chapter 11
Sheet Metal: Modify Sheet Metal Parts	Demonstrate the creation of a corner seam.	Chapter 11
	Demonstrate the creation of a punch tool.	Chapter 11
	Demonstrate the creation of a cut across a bend.	Chapter 11
Sheet Metal: Flat Pattern	Demonstrate how to create a flat pattern.	Chapter 11
	Demonstrate how to insert a flat pattern in a drawing.	Chapter 11
	Demonstrate how to export a flat pattern.	Chapter 11
Visualization: Create Rendered Images	Describe the process to activate Inventor Studio.	Bonus Chapter 4
	Demonstrate how to create a new camera.	Bonus Chapter 4
	Demonstrate how to create a rendered image.	Bonus Chapter 4
Visualization: Animate an Assembly	Demonstrate how to create a new animation.	Bonus Chapter 4
	Demonstrate how to create an animation by animating a camera.	Bonus Chapter 4
	Demonstrate how to create an animation by animating a constraint.	Bonus Chapter 4
	Demonstrate how to create an animation by animating a fade.	Bonus Chapter 4

TABLE A.2   Autodesk Certified Professional exam sections and objectives

Topic	Learning Objective	Chapter
Advanced Modeling	Create a 3D path using the Intersection Curve and the Project to Surface commands.	Chapters 3, 6, 7, Bonus Chapter 3
	Create a loft feature.	
	Create a multibody part.	
	Create a part using surfaces.	
	Create a sweep feature.	
	Create an iPart.	
	Create and constrain sketch blocks.	
	Use iLogic.	
	Emboss text and a profile.	
Assembly Modeling	Apply and use assembly constraints.	Chapters 5, 9, 12, Bonus Chapter 2
	Create a level of detail.	
	Create a part in the context of an assembly.	
	Describe and use Shrinkwrap.	
	Create a positional representation.	
	Create components using the Design Accelerator commands.	
	Modify a bill of materials.	
	Find minimum distance between parts and components.	
	Use the frame generator command.	
Drawing	Create and edit dimensions in a drawing.	Chapters 4, 8, Bonus Chapter 1
	Edit a section view.	
	Modify a style in a drawing.	
	Edit a hole table.	
	Modify a parts list.	
	Edit a base and projected views.	

Part Modeling	Create a pattern of features.	Chapters 2, 6, 7
	Create a shell feature.	
	Create extrude features.	
	Create fillet features.	
	Create hole features.	
	Create revolve features.	
	Create work features.	
	Use the Project Geometry and Project Cut Edges commands.	
Presentation Files	Animate a presentation file.	Bonus Chapter 4
Project Files	Control a project file.	Chapter 1
Sheet Metal	Create flanges.	Chapter 11
	Annotate a sheet metal part in a drawing.	
	Create and edit a sheet metal flat pattern.	
	Describe sheet metal features.	
Sketching	Create dynamic input dimensions.	Chapters 2, 6, 7, 11
	Use sketch constraints.	
User Interface	Identify how to use visual styles to control the appearance of a model.	Chapter 1
Weldments	Create a weldment.	Chapter 13

INDEX

Note to the Reader: Throughout this index **boldfaced** page numbers indicate primary discussions of a topic. *Italicized* page numbers indicate illustrations.